# MTEL Earth Science
## 14 Teacher Certification Exam

**By:** Sharon Wynne, M.S
Southern Connecticut State University

**XAMonline, INC.**
Boston

**Copyright © 2007 XAMonline, Inc.**
All rights reserved. No part of the material protected by this copyright notice may be reproduced or utilized in any form or by any means, electronic or mechanical, including photocopying, recording or by any information storage and retrievable system, without written permission from the copyright holder.

To obtain permission(s) to use the material from this work for any purpose including workshops or seminars, please submit a written request to:

XAMonline, Inc.
21 Orient Ave.
Melrose, MA 02176
Toll Free 1-800-509-4128
Email: info@xamonline.com
Web www.xamonline.com
Fax: 1-781-662-9268

Library of Congress Cataloging-in-Publication Data

Wynne, Sharon A.
 Earth Science 14: Teacher Certification / Sharon A. Wynne. -2$^{nd}$ ed.
 ISBN 978-1-58197-683-0
  1. Earth Science 14.          2. Study Guides.        3. MTEL
  4. Teachers' Certification & Licensure.    5. Careers

**Disclaimer:**
The opinions expressed in this publication are the sole works of XAMonline and were created independently from the National Education Association, Educational Testing Service, or any State Department of Education, National Evaluation Systems or other testing affiliates.

Between the time of publication and printing, state specific standards as well as testing formats and website information may change that is not included in part or in whole within this product. Sample test questions are developed by XAMonline and reflect similar content as on real tests; however, they are not former tests. XAMonline assembles content that aligns with state standards but makes no claims nor guarantees teacher candidates a passing score. Numerical scores are determined by testing companies such as NES or ETS and then are compared with individual state standards. A passing score varies from state to state.

**Printed in the United States of America** œ-1

# MTEL: Earth Science 14
ISBN: 978-1-58197-683-0

TEACHER CERTIFICATION STUDY GUIDE

# Table of Contents

**SUBAREA I.**                                             **SCIENTIFIC INQUIRY**

**Competency 1.0**   Understand types and uses of natural resources, the effects of human activities on the environment, and the need to preserve the environmental integrity of the Earth's ecosystems ...................................... 1

Skill 1.1   Understand the classification, uses, and importance of natural resources and methods of locating and obtaining natural resources .................................................. 1

Skill 1.2   Identify the positive and negative effects of human activities on Earth's environment (e.g., reclamation of strip mines, ocean dumping) ........................................... 5

Skill 1.3   Describe strategies for dealing with environmental problems................................................................................ 12

**Competency 2.0**   Understand the nature of scientific inquiry, the role of observation and experimentation in science, and the relationships between Earth science, technology, and other fields of knowledge...................................... 15

Skill 2.1   Identify the processing by which new scientific knowledge and hypotheses are generated and rejected .... 15

Skill 2.2   Understand ethical issues related to Earth science (e.g., accurately reporting experimental results .................. 18

Skill 2.3   Evaluate the appropriateness of a specified experimental design to test a given hypothesis in Earth science ............. 18

Skill 2.4   Understand the role of communication among scientists in promoting scientific progress. ........................................ 19

Skill 2.5   Identify the similarities and differences between science and technology (e.g., science as investigating the natural world, technology as solving human adaptation problems) ............................................................................. 20

**Competency 3.0**   Understand the processes of gathering, organizing, reporting, and interpreting scientific data in the context of Earth science investigations. .......................... 23

Skill 3.1   Evaluate the appropriateness of a given method or procedure for collecting data for a specified purpose ........ 23

# TEACHER CERTIFICATION STUDY GUIDE

| Skill 3.2 | Organize and report given experimental data using appropriate and effective graphic representation (e.g., graphs, tables, diagrams) .................................................. 23 |
|---|---|
| Skill 3.3 | Identify the procedures and criteria for formally reporting experimental procedures and data to the scientific community .................................................................. 23 |
| Skill 3.4 | Identify relationships between factors (e.g. inverse, direct, linear) as indicated by experimental data ............... 24 |

**Competency 4.0** **Understand how to create, use, and interpret physical and mathematical models (e.g. maps, charts, graphs, diagrams) commonly used in Earth science ....... 26**

| Skill 4.1 | Describe alternative models for conveying given information from Earth science ........................................... 26 |
|---|---|
| Skill 4.2 | Describe methods by which given physical and graphic models are created ................................................ 27 |
| Skill 4.3 | Classify different types of maps (e.g. topographic, geologic) used in Earth science and analyze the information conveyed by that type of map ........................ 29 |
| Skill 4.4 | Interpret diagrams relating to Earth science (e.g. crustal movements) ..................................................... 31 |

**Competency 5.0** **Understand proper and safe use of equipment and materials used in Earth science investigations ........................................................................................ 33**

| Skill 5.1 | Identify the operating principles for various instruments ................................................................................ 33 |
|---|---|
| Skill 5.2 | Describe basic safety procedures in a laboratory or field situation ( e.g. wearing safety glasses) ................... 34 |
| Skill 5.3 | Apply proper procedures for dealing with given accidents and injuries in the Earth science laboratory or in the field ................................................................................ 35 |

TEACHER CERTIFICATION STUDY GUIDE

| SUBAREA II. | | ASTRONOMY |
|---|---|---|

| Competency 6.0 | Understand the physical characteristics and motions of the Earth as well as the evidence of and methods used to determine these characteristics and motions ............................................................................ 37 |
|---|---|
| Skill 6.1 | Describe historical methods used to study the characteristics and motions of the Earth (e.g., Foucault pendulum)................................... 37 |
| Skill 6.2 | Identify the physical characteristics of the Earth (e.g., diameter, tilt of axis, distance from the Sun) and how they can be determined ............................................. 39 |
| Skill 6.3 | Describe the consequences of the Earth's motions (e.g., length of day, change in seasons, length of year) ..... 40 |
| Competency 7.0 | Understand the structure, composition, and features of the Sun (including its production and transmission of energy) and the importance of the Sun to Earth processes ............................................................. 41 |
| Skill 7.1 | Describe the methods by which the Sun's diameter, surface temperature, and chemical composition are determined ................................................................. 41 |
| Skill 7.2 | Identify the layers of the Sun ............................................ 43 |
| Skill 7.3 | Describe the Sun's source of energy (fusion reaction) ....... 44 |
| Skill 7.4 | Describe the solar surface as the immediate source of energy fro the Earth's surface ........................................... 44 |
| Skill 7.5 | Understand the sunspot cycle and its possible effects on the Earth's climate ...................................................... 45 |
| Competency 8.0 | Understand the properties, features, and movements of the Earth's moon; the interactions among the Earth, Moon, and Sun (including phases, tides, and eclipses); and the role of technology and exploration in obtaining knowledge about the Earth, Moon, and Sun ....... 47 |
| Skill 8.1 | Relate surface features (e.g., maria, craters, mountains) of the Earth's moon to events in the history of the Moon.... 47 |

EARTH SCIENCE

| | | |
|---|---|---|
| Skill 8.2 | Describe the relationship between the height of ocean tides and the relative positions of the Earth, Moon, and Sun | 47 |
| Skill 8.3 | Describe the relationships between the phases of the Moon and the relative positions of the Earth, Moon, and Sun | 48 |
| Skill 8.4 | Explain how the lunar exploration program has added to our knowledge of the Earth Moon system | 49 |
| **Competency 9.0** | **Understand the scale and organization of the solar system, the role of gravity in the solar system, characteristics of the bodies within the solar system, and physical and mathematical models that describe these objects and their real and apparent motions** | **50** |
| Skill 9.1 | Describe the characteristics (e.g., size, density, surface temperature) of the planets | 50 |
| Skill 9.2 | Identify the relative sizes, distances, tilts, and positions of the planets | 53 |
| Skill 9.3 | Identify the position of the planets on the ecliptic | 54 |
| Skill 9.4 | Describe the origin and properties of comets and meteors | 54 |
| Skill 9.5 | Use the apparent motion of celestial objects to infer solar system models (i.e., geocentric, heliocentric). | 55 |
| Skill 9.6 | Apply Kepler's laws to describe and predict the motions of the planets. | 57 |
| **Competency 10.0** | **Understand stars, their motions and life cycles, and the methods and technology used to study them** | **58** |
| Skill 10.1 | Compare types of telescopes (e.g., optical, radio, infrared, ultraviolet) and the ways in which they are used to acquire information on star characteristics | 58 |
| Skill 10.2 | Identify methods and uses of spectroscopy | 61 |
| Skill 10.3 | Identify types of stars (e.g., pulsars, Cepheid variables) and their characteristics | 62 |

| | | |
|---|---|---|
| Skill 10.4 | Use the H-R diagram to analyze the life cycles of stars | 64 |
| Skill 10.5 | Analyze stellar life cycles to understand the formation and initial development of the solar system | 65 |

**Competency 11.0 Understand evidence regarding the size, structure, scale, and motions of the universe, the Milky Way galaxy, and the solar system ............................. 69**

| | | |
|---|---|---|
| Skill 11.1 | Describe the evidence regarding the location of the solar system within the Milky Way galaxy | 69 |
| Skill 11.2 | Identify historical methods of inferring the size, structure, and motions of the galaxy and the solar system (e.g., star observations and counts) | 70 |
| Skill 11.3 | Describe the evidence for and interpretations of an expanding universe (e.g., red shift and the Doppler effect) | 72 |
| Skill 11.4 | Analyze types of evidence used to infer scales and relative motions of the solar system, the Milky Way galaxy, the universe (e.g., in relation to relative size and distance) | 72 |

**SUBAREA III.**                                                                               **METEOROLOGY**

**Competency 12.0 Understand the composition, structure, and properties of the Earth's atmosphere and the mechanisms and effects of energy transfer involving the Earth-atmosphere system ............................................. 75**

| | | |
|---|---|---|
| Skill 12.1 | Identify the properties (e.g., density, composition, temperature) of the atmosphere from the Earth's surface through the thermosphere and the significance of changes in these properties | 75 |
| Skill 12.2 | Analyze ho various wavelengths of solar radiation (e.g., ultraviolet, visible light, infrared) are affected as the radiation enters and passes through the atmosphere and is absorbed by and radiated from the Earth's surface | 77 |
| Skill 12.3 | Identify the processes by which energy in transferred to and within the atmosphere (e.g., radiation, convection, conduction) | 79 |

# TEACHER CERTIFICATION STUDY GUIDE

Skill 12.4 Analyze global wind patterns in terms of latitudinal variations in insolation and the Corilois effect. ................... 80

**Competency 13.0 Understand the properties of water, conditions in the atmosphere that favor phase changes, and the energy relationships among phase changes, cloud formation, and precipitation ................................................... 83**

Skill 13.1 Relate the physical properties of water (e.g., high specific heat, surface tension) to the chemical structure and properties of water molecules ...................... 83

Skill 13.2 Describe the energy changes involved in the transition between phases of water (i.e., latent heat) ......... 84

Skill 13.3 Describe the atmospheric conditions under which fog and clouds with various characteristics form (e.g., adiabatic temperature changes, dew-point, atmospheric stability) ................................................................. 85

Skill 13.4 Describe the conditions under which precipitation forms ................................................................. 89

Skill 13.5 Predict the type of precipitation that will fall to the Earth's surface under given conditions ............................................ 90

**Competency 14.0 Understand characteristics of broad-scale weather systems and local weather, the relationship between them, and the methods and instruments used to collect weather data ........................................................ 93**

Skill 14.1 Identify the types and characteristics of air masses, their movements, the kinds of fronts that form between air masses ................................................................. 93

Skill 14.2 Identify the horizontal and vertical movements of air in high- and low-pressure areas ........................................ 94

Skill 14.3 Describe the use of weather instruments (e.g., thermometer, psychrometer) for collecting given types of weather data ................................................................. 96

**Competency 15.0 Understand weather maps, how they are prepared, and how they are used. ............................................... 98**

Skill 15.1 Interpret symbols used on weather maps ........................... 98

**EARTH SCIENCE**

| Skill 15.2 | Describe the methods used to generate weather maps ................................................................ 102 |
|---|---|
| Skill 15.3 | Infer recent weather in a given location based on one or more weather maps ................................. 110 |
| Skill 15.4 | Predict future weather in a given location based on one or more weather maps ............................. 114 |

**Competency 16.0 Understand the principles and technology of weather forecasting and the effects of weather and weather forecasting on humans ......................................... 115**

| Skill 16.1 | Describe the use of weather models in forecasting ............ 115 |
|---|---|
| Skill 16.2 | Identify the role of computers and satellite photographs in generating weather forecasts. ........................................ 115 |
| Skill 16.3 | Describe types of hazardous weather ............................... 118 |
| Skill 16.4 | Identify types and functions of weather precautions ........... 124 |
| Skill 16.5 | Explain the role of the National Weather Service in issuing weather alerts.......................................................... 126 |

**Competency 17.0 Understand the locations and characteristics of the Earth's major climatic regions and analyze factors that affect local climate and the relationship between weather and climate ............................................. 129**

| Skill 17.1 | Infer the climatic zone in which a given area is located based on temperature and precipitation data ..................... 129 |
|---|---|
| Skill 17.2 | Identify the factors that affect the climate in a given region (e.g., insolation, wind patterns, topography)............ 131 |
| Skill 17.3 | Describe the relationship between the climate of a region and its weather ......................................................... 133 |

**Competency 18.0 Understand the effects of human activities and natural processes on the atmosphere, theories about the long-range effects of human activities on global climate, and methods of controlling and minimizing those effects ......................................................................... 134**

| Skill 18.1 | Identify common air pollutants and their sources and effects............................................................................. 134 |
|---|---|

| Skill 18.2 | Describe pollutants and atmospheric chemical reactions involving these pollutants ................................................. 137 |
|---|---|
| Skill 18.3 | Describe factors that affect local air pollutant concentrations (e.g., population density)............................ 138 |
| Skill 18.4 | Analyze the theory of global warming due to increased levels of atmospheric carbon dioxide from the burning of fossil fuels.................................................... 139 |

## SUBAREA IV.                                              GEOLOGY AND OCEANOGRAPHY

| Competency 19.0 | **Understand the processes of mineral and rock formation, the characteristics of different types of minerals and rocks, and the methods used to identify and classify them................................................ 144** |
|---|---|
| Skill 19.1 | Use classification schemes (e.g., Mohs' scale of hardness, crystal form, chemical composition) to identify common rock-forming minerals ............................................ 144 |
| Skill 19.2 | Identify the processes by which different kinds of rocks are formed ...................................................................... 148 |
| Skill 19.3 | Classify a given rock as sedimentary, igneous, metamorphic ................................................................... 154 |
| Competency 20.0 | **Understand the structure of the earth, the constructional forces that have shaped its surface, theories and evidence of crustal movements, and the effects of crustal movements on landscape ....................... 158** |
| Skill 20.1 | Use seismic waves to infer the Earth's internal structure ........................................................................... 158 |
| Skill 20.2 | Use the theory of dynamic equilibrium (e.g., isostasy) to explain landscape development ...................................... 159 |
| Skill 20.3 | Describe the evidence for continental drift and seafloor spreading ......................................................................... 160 |
| Skill 20.4 | Apply the theory of plate tectonics to explain landscape development and geologic phenomena (e.g., volcanism, earthquakes) and to predict future movements of land masses............................................... 162 |

# TEACHER CERTIFICATION STUDY GUIDE

**Competency 21.0** **Understand erosional-depositional processes that change the Earth's surface (e.g., weathering, erosion) and the relationship between these processes and landscape development ......................................... 164**

Skill 21.1 Describe the processes of mechanical, chemical, and biological weather and factors that affect the rate at which rocks weather and soils are produced. ................................................................................ 164

Skill 21.2 Identify the processes of erosion by various agents (e.g., wind, water, glaciers) and factors that affect erosion rates and patterns. ...................................... 167

Skill 21.3 Explain the processes by which given landscape features (e.g., eskers, moraines) are formed. .................... 170

Skill 21.4 Describe the effects of glaciation on the Massachusetts landscape. ............................................................................ 172

**Competency 22.0** **Understand characteristics of the major geologic time divisions and theories and supporting evidence of the Earth's geologic history ......................................... 173**

Skill 22.1 Identify the conditions and characteristic fossils of the various geologic periods. ............................................. 173

Skill 22.2 Apply the laws and principles of geology (e.g., law of original horizontality, law of superposition) to interpret diagrams of rock strata. ...................................................... 175

Skill 22.3 Identify the principles, applications, and limits of radioactive dating. .................................................................... 178

Skill 22.4 Use paleontological information to infer to geologic history of a given area. ...................................................... 179

**Competency 23.0** **Understand the hydrologic cycle and the processes by which water moves on and beneath the Earth's surface, and use this knowledge to analyze local weather budgets ................................................... 181**

Skill 23.1 Analyze a cross-sectional diagram of a water table and surrounding rock strata to predict movement of groundwater (e.g., aquifers, gradient). .......................... 181

**EARTH SCIENCE**

| | | |
|---|---|---|
| Skill 23.2 | Describe the effects of various factors (e.g., vegetation, gradient, rock strata) on components of a local water budget. | 182 |

**Competency 24.0 Understand ocean water and its movements ....... 185**

| | | |
|---|---|---|
| Skill 24.1 | Describe the circulation patterns in the oceans and factors that influence these patterns (e.g., temperature variations, wind systems, Coriolis effect). | 185 |
| Skill 24.2 | Describe the types, causes, and effects of tidal and wave motions of ocean water. | 188 |

**Competency 25.0 Understand the structure and topography of the ocean basin. ................................................................. 190**

| | | |
|---|---|---|
| Skill 25.1 | Identify ocean zones (e.g., littoral, benthic) in terms of their physical characteristics. | 190 |
| Skill 25.2 | Describe the major structural features of the ocean floor. | 191 |
| Skill 25.3 | Identify the factors involved in changing the structure of the ocean floor. | 193 |

**Competency 26.0 Understand marine life and the marine habitat ................................................................................ 195**

| | | |
|---|---|---|
| Skill 26.1 | Identify the characteristics and major groups of marine plants and animals. | 195 |
| Skill 26.2 | Describe the zonation of marine plants and animals. | 197 |
| Skill 26.3 | Describe the relationships between marine organisms and the marine environment. | 199 |

**Sample Test** ........................................................................................ 201

**Answer Key** ....................................................................................... 212

**Rationales with Sample Questions** ................................................ 213

# TEACHER CERTIFICATION STUDY GUIDE

## Great Study and Testing Tips!

*What* to study in order to prepare for the subject assessments is the focus of this study guide but equally important is *how* you study.

You can increase your chances of truly mastering the information by taking some simple, but effective steps.

## Study Tips:

**1. Some foods aid the learning process.** Foods such as milk, nuts, seeds, rice, and oats help your study efforts by releasing natural memory enhancers called CCKs (*cholecystokinin*) composed of *tryptopha*n, *choline*, and *phenylalanine*. All of these chemicals enhance the neurotransmitters associated with memory. Before studying, try a light, protein-rich meal of eggs, turkey, and fish. All of these foods release the memory enhancing chemicals. The better the connections, the more you comprehend.

Likewise, before you take a test, stick to a light snack of energy boosting and relaxing foods. A glass of milk, a piece of fruit, or some peanuts all release various memory-boosting chemicals and help you to relax and focus on the subject at hand.

**2. Learn to take great notes.** A by-product of our modern culture is that we have grown accustomed to getting our information in short doses (i.e. TV news sound bites or USA Today style newspaper articles.)

Consequently, we've subconsciously trained ourselves to assimilate information better in neat little packages. If your notes are scrawled all over the paper, it fragments the flow of the information. Strive for clarity. Newspapers use a standard format to achieve clarity. Your notes can be much clearer through use of proper formatting. A very effective format is called the *"Cornell Method."*

> Take a sheet of loose-leaf lined notebook paper and draw a line all the way down the paper about 1-2" from the left-hand edge.
>
> Draw another line across the width of the paper about 1-2" up from the bottom. Repeat this process on the reverse side of the page.

Look at the highly effective result. You have ample room for notes, a left hand margin for special emphasis items or inserting supplementary data from the textbook, a large area at the bottom for a brief summary, and a little rectangular space for just about anything you want.

# TEACHER CERTIFICATION STUDY GUIDE

**3. Get the concept then the details.** Too often we focus on the details and don't gather an understanding of the concept. However, if you simply memorize only dates, places, or names, you may well miss the whole point of the subject.

A key way to understand things is to put them in your own words. If you are working from a textbook, automatically summarize each paragraph in your mind. If you are outlining text, don't simply copy the author's words.

*Rephrase* them in your own words. You remember your own thoughts and words much better than someone else's, and subconsciously tend to associate the important details to the core concepts.

**4. Ask Why?** Pull apart written material paragraph by paragraph and don't forget the captions under the illustrations.

Example: If the heading is "Stream Erosion", flip it around to read "Why do streams erode?" Then answer the questions.

If you train your mind to think in a series of questions and answers, not only will you learn more, but it also helps to lessen the test anxiety because you are used to answering questions.

**5. Read for reinforcement and future needs.** Even if you only have 10 minutes, put your notes or a book in your hand. Your mind is similar to a computer; you have to input data in order to have it processed. *By reading, you are creating the neural connections for future retrieval.* The more times you read something, the more you reinforce the learning of ideas.

Even if you don't fully understand something on the first pass, *your mind stores much of the material for later recall.*

**6. Relax to learn so go into exile.** Our bodies respond to an inner clock called biorhythms. Burning the midnight oil works well for some people, but not everyone.

If possible, set aside a particular place to study that is free of distractions. Shut off the television, cell phone, and pager and exile your friends and family during your study period.

If you really are bothered by silence, try background music. Light classical music at a low volume has been shown to aid in concentration over other types. Music that evokes pleasant emotions without lyrics is highly suggested. Try just about anything by Mozart. It relaxes you.

**EARTH SCIENCE**

**7. Use arrows not highlighters.** At best, it's difficult to read a page full of yellow, pink, blue, and green streaks. Try staring at a neon sign for a while and you'll soon see that the horde of colors obscure the message.

A quick note, a brief dash of color, an underline, and an arrow pointing to a particular passage is much clearer than a horde of highlighted words.

**8. Budget your study time.** Although you shouldn't ignore any of the material, *allocate your available study time in the same ratio that topics may appear on the test.*

# TEACHER CERTIFICATION STUDY GUIDE

## Testing Tips:

**1. Get smart, play dumb. Don't read anything into the question.** Don't make an assumption that the test writer is looking for something else than what is asked. Stick to the question as written and don't read extra things into it.

**2. Read the question and all the choices *twice* before answering the question.** You may miss something by not carefully reading, and then re-reading both the question and the answers.

If you really don't have a clue as to the right answer, leave it blank on the first time through. Go on to the other questions, as they may provide a clue as to how to answer the skipped questions.

If later on, you still can't answer the skipped ones . . . **Guess.** The only penalty for guessing is that you *might* get it wrong. Only one thing is certain; if you don't put anything down, you will get it wrong!

**3. Turn the question into a statement.** Look at the way the questions are worded. The syntax of the question usually provides a clue. Does it seem more familiar as a statement rather than as a question? Does it sound strange?

By turning a question into a statement, you may be able to spot if an answer sounds right, and it may also trigger memories of material you have read.

**4. Look for hidden clues.** It's actually very difficult to compose multiple-foil (choice) questions without giving away part of the answer in the options presented.

In most multiple-choice questions you can often readily eliminate one or two of the potential answers. This leaves you with only two real possibilities and automatically your odds go to Fifty-Fifty for very little work.

**5. Trust your instincts.** For every fact that you have read, you subconsciously retain something of that knowledge. On questions that you aren't really certain about, go with your basic instincts. **Your first impression on how to answer a question is usually correct.**

**6. Mark your answers directly on the test booklet.** Don't bother trying to fill in the optical scan sheet on the first pass through the test.

*Just be very careful not to miss-mark your answers when you eventually transcribe them to the scan sheet.*

**7. Watch the clock!** You have a set amount of time to answer the questions. Don't get bogged down trying to answer a single question at the expense of 10 questions you can more readily answer.

**EARTH SCIENCE**

| SUBAREA I. | SCIENTIFIC INQUIRY |

**COMPETENCY 1.0** Understand types and uses of natural resources, the effects of human activities on the environment, and the need to preserve the environmental integrity of the earth's ecosystems.

**SKILL 1.1 Understand the classification, uses, and importance of natural resources and methods of locating and obtaining natural resources**

**Natural Resources:** naturally created commodities critically important or necessary to human life and civilization. The term natural resource can also include the total quantity of a given resource commodity on Earth, discovered and undiscovered.

A major source of contention in our modern society centers on the proper use and conservation of our natural resources. Although most people automatically think of coal, oil, iron, and other minerals when they think of natural resources, the definition also includes other often overlooked resources such as forests, soil, water, air, and land.

Our natural resources are classified into two broad categories: **Renewable Resources** and **Nonrenewable Resources**.

**Renewable Resource:** a resource that is capable of replenishment or regeneration on a human time scale. Example: Forests and water.

**Nonrenewable Resource:** a resource that, once exhausted, is not capable of replenishment or regeneration on a human time scale. Example: Petroleum and minerals.

With the exception of certain commodities (gold, silver, and salt), humans have historically given little thought to the concept of managing resources. The supply of every conceivable resource seemed boundless. Vast forests covered large parts of the planet, ensuring a seemingly ample supply of wood for heating and construction. Ore deposits, although less accessible than the forests, were nonetheless plentiful.

At first, our natural **reserves**, the quantity of a resource material that has been discovered and is economically or technologically recoverable, kept pace with the increased demand. However, as the population increased and civilization advanced, becoming increasingly technologically dependent, the demand on our natural resources soared. More important, the per capita consumption of resources has dramatically increased. With an increased standard of living comes the specter of increased individual consumption and strong dependence on machinery and manufactured goods. **Renewable Resource Concerns**

We have a watery planet. Unfortunately, a large percentage (97%) of the water is not fit for human consumption or agricultural use due to high salinity.

Plants and animals (including humans) require water for survival. In fact, statistics show that every person in the United States uses 300 liters of water, and when industrial uses are included, that number soars to roughly 5,000 liters per day, per person.

Pollution poses the severest threat to the water supply. Organic wastes (sewage) are produced by both humans and animals. Left untreated, these wastes, and the wastes from food treatment plants, can enter the waterways and upset the ecological balance. As the wastes decay, they consume oxygen in the water, depriving aquatic life of oxygen, or causing algae blooms which further deplete the oxygen supply, eventually turning some water anoxic.

Another danger to ecology is the poisoning of the food chain through pesticides and fertilizers, or with high concentrations of heavy metals carried into the water supply through runoff from farmlands, factories, and mine tailings.

Pollution also affects our air. The uncontrolled burning of fossil fuel hydrocarbons and high-sulphur content coals pose severe health risks, especially to the very young and very old. Smog alerts are routine in many of the major metropolitan areas, and in Mexico City, air pollution is reaching a critical level.

Forestry management is another area of concern. As our population grows, the demand for lumber and wood products has grown exponentially. Increased urbanization has claimed once vast tracts of forests, replacing them with concrete paving and closely packed structures.

This same drive to urbanization also affects our soil. Arable farmland is shrinking as the press to develop home and commercial sites increases. Of the approximately 15 billion hectares of dry land on the Earth, only 2 billion is suitable for agriculture. If the same land is used year after year, there is a definite danger of soil exhaustion as vital nutrients are depleted.

Farmland is not the only victim of urbanization. Grazing lands for our cattle and other domesticated animals are also shrinking, and as a consequence many of the remaining areas are being overgrazed. The danger of overgrazing lays in the non-availability of sufficient pasturage for the animals and the loss of top cover for the soil, which is then left vulnerable to erosion.

## Nonrenewable Resource Concerns

The key focus in nonrenewable resources is the increasing demand for energy. The concern with regard to nonrenewable resources is that once they are depleted, they are permanently gone.

Despite a finite supply of fossil fuels and radioactive fuels such as uranium, the demand for energy continues to increase at a high rate. At our present rate of consumption, there are only 28 years of petroleum reserves left, and uranium reserves are estimated at depletion in 40 years.

To try and alleviate this predictable energy gap, scientists are exploring new methods of recovering additional fuels from once economically unfeasible sites, and researching alternative energy sources.

## Alternative Energy Resources

Research efforts into alternative energy sources are directed at producing viable renewable energy sources.

**Hydroelectric Power:** power produced from falling water. This is not a new idea as waterwheels have been in use for centuries. The drawback to this energy source lays in the availability of suitable locations for dams, and the expense of their construction.

**Wind Power:** Windmills are also another ancient technology being revisited by engineers. However, the **wind generators** produce very little electricity for the expense involved, and the suitable locations (steady, high winds) for **windfields** are limited.

**Tidal Power:** Another concept in use in some areas of the world is generating electricity by deflecting and diverting strong tidal currents through offshore turbines that drive electric generators. Again, the presence of proper conditions is necessary (strong tidal power), and suitable locations are limited.

**Geothermal Energy:** In some areas of the world, such as New Zealand, Iceland, and Italy, energy is produced from hot igneous rocks within the Earth. Rainwater percolates through porous strata near an active magma chamber and flashes into steam. Some of the steam returns to the surface through natural fissures or is extracted through drilled vents. The steam is captured and routed to turbine powered electrical generators to produce geothermal power.

An additional use of geothermal energy is that the steam may also be used to directly heat buildings. Example: Reykjavik, Iceland uses the captured steam to directly heat their buildings. The limitations of this alternative energy source are obvious; the majority of metropolitan locations are not situated near active magma chambers. However, New Zealand does manage to gather enough power to meet approximately 5% of their electrical needs.

**Solar Energy:** Solar power can be utilized directly as a source of heat or to produce electricity. The most common use is to heat water. An array of dark colored piping is placed on the roof of a structure and as water circulates through the piping it is heated by the Sun.

**Solar cells** produce electricity from solar radiation. Photons striking the junction between two semi-conductors (usually selenium) induce an electrical current that is stored in batteries. Although this source of power is pollution free, there are two main limitations; first the production of power is limited by the distribution and periods of insolation, and atmospheric conditions can easily interfere with collection efforts (i.e. winter months, cloud cover, pollution, and storms).

Secondly, the solar cells individually produce very small amounts of electricity (trickle charges) and must be arrayed in very large banks. Example: A solar power plant with a capacity of 100 MW would cover a surface area of approximately 4 $km^2$.

Solar cells have been used successfully in outer space where atmospheric conditions and cell size restraints are of less concern. Spacecraft and satellites use solar cells to charge batteries that provide electrical power for communications equipment and operating power.

**Biomass:** Plant and animal wastes (decaying or decayed) can be burned to produce heat for steam turbine electrical generators. In most highly developed countries, the biomass is first converted to either methane gas (given off by decaying biomass) or alcohol, but in some underdeveloped countries, the biomass is still burned directly as a fuel source. Example: For centuries, **Peat Bogs** were exploited as a traditional source of home heating and cooking fuel.

**Fusion Power:** Although the technology does not currently exist, researchers are actively pursuing the means to make fusion power a reality. Unlike *Fission*, the other form of nuclear energy currently in use, *Fusion* does not rely on splitting the atoms of uranium or other potentially deadly radioactive elements. Instead, fusion energy mimics the same process that produces the energy of the Sun.

Energy is produced when small atomic nuclei fuse together to form new atoms. In a fusion reaction, two isotopes of hydrogen, deuterium, and tritium combine to make helium.

The most significant advantage offered by fusion power as compared to fission power is that no dangerous radioactive isotopes are produced. The reaction produces only harmless helium that easily diffuses into the atmosphere and escapes into outer space. Additionally, the elements required for a fusion reaction are abundant on Earth (i.e. deuterium and tritium are extracted from seawater), and readily renew themselves through natural processes.

**SKILL 1.2    Identify the positive and negative effects of human activities on earth's environment (e.g., reclamation of strip mines, ocean dumping).**

**Positive Effects – Reclamation of Strip Mines**

Mining is the process by which minerals are extracted from the Earth. These minerals may often include coal, limestone, gold, silver, and many other metals. Mining causes the disturbance of land and ecosystems. Mine reclamation is the process by which mined land is restored to a useful state, such as a productive ecosystem or industrial or municipal land. Mine reclamation has become a regular part of modern mining industry, and improves water and air quality in abandoned mine areas. Reclaimed sites may function as pasture areas, hayland, recreational areas, wild life habitat, and wetlands.

Mine reclamation techniques stabilize land surfaces against water and wind erosion using material placement and capping. The final step in mine reclamation is often the replacement of topsoil and its revegetation with suitable plant species. Revegetation techniques include hydroseeding, a process commonly used for large-scale or hillside properties in which grass seed is sown in a stream of water aimed at the ground, as well as native seed drilling techniques, through which seeds can be sown in well-spaced rows at specific depths. Tree planting is another important part of mine reclamation. Trees are generally planted in low densities to allow for natural propagation. Tree seeds can be pelletalized to prevent excessive movement by the wind. This method proves particularly effective on rocky, barren slopes.

Tailing basins are a common characteristic of mined areas. These depressed areas contain a finely ground concentrate that is the byproduct of smelting operations conducted at mine sites. As the content levels of these basins increase, wind erosion leads to contamination of electrometallurgical refining processes, problems for residents of nearby towns, and machinery wear. To revegetate such areas and decrease erosion, seeding is established on portions of the basin closest to prevailing winds in order to minimize damage of young plants by eroding soil. Agricultural limestone can be applied to soil approximately six weeks before seeding to raise pH, and nitrogenous fertilizers are often used to encourage growth of young plants. When planting grasses, companion crops can be employed to create shading and reduce the dry effect of wind. To deal with drainage problems found in tailing basins, wetlands are often constructed downstream of the basins. These wetlands are capable of filtering and removing contaminants through biological and chemical processes.

**Negative Effects – Ocean Dumping**

In 1972, the Marine Protection, Research, and Sanctuaries Act was issued to "prevent or strictly limit the dumping into ocean waters of any material that would adversely affect human health, welfare, or amenities, or the marine environment, ecological systems, or economic potentialities." Prior to this act, the dumping of industrial, nuclear, and other waste products into the ocean was common practice. For example, a 1970 Report to the President from the Council on Environmental Quality concluded that 38 million tons of dredged material (34% of which was polluted), 4.5 million tons of industrial waste, 4.5 million tons of sewage sludge containing heavy metals, and 0.5 million tons of construction and demolition debris was disposed of in ocean waters in 1968.

Currently, illegal ocean dumping is widespread. The majority of material dumped at sea is generated by dredging and amounts to hundreds of millions of tons each year. Approximately 10% of this material contains pollutants from shipping, industrial and municipal discharges, and land run off. These pollutants include heavy metals, such as cadmium, mercury, and chromium, hydrocarbons such as oil, pesticides, and nutrients such as nitrogen and phosphorous.

Ocean dumping may result in the rapid and substantial degradation of marine water quality and ecosystems. Toxic materials disposed of at sea may cause acute or chronic toxic effects on marine organisms and potentially contaminate human food sources. Dumping may also cause drastic decreases in ocean water oxygen concentrations, resulting in dead zones. When one habitat or species is affected by ocean pollution, the survival of organisms that rely on that niche is compromised as well, resulting in massive loss of life and even extinction. For example, sea turtles, manatees, fish, shrimp and crabs rely on sea grass for survival. This plant is particularly sensitive to pollution, and its numbers are decreasing drastically. Loss of sea grass affects the survival of animals such as sea lions, which in turn causes sea urchins, the sea lions' prey, to increase in number, upsetting the delicate ecosystem of the ocean.

Another danger to ocean ecology is the poisoning of the food chain through pesticides and fertilizers, or with high concentrations of heavy metals carried into the water supply through runoff from farmlands, factories, and mine tailings.

**Landfills:** Landfill toxins that leech into soil and groundwater are the primary cause. However, this is preventable and new federal standards require landfills to have a protective plastic barrier to prevent leaching.

**Leaking underground storage tanks:** This used to be a major source of pollution, but the federal government required the removal of old tanks and replacement with double-hulled, leak alarmed tanks.

**Pipelines:** Not a major source of concern since new construction techniques generally preclude routine leakage.

**Contaminated Wells:** New methods of controlling contamination by hydrocarbons (oil, gas, old tires, etc.) have shown promise.

**Biomediation** combines bacteria and nutrients pumped into the groundwater to consume the contamination. However, it is a long, slow process of over 30 days.

**Road de-icing chemicals**: This mostly affects areas that receive a significant amount of snow and ice. The chemicals leach into the water table. The only method of dealing with this problem is to use improved chemicals and use less of them.

**Fertilizers:** Farm chemicals are not a major threat because they are designed to be biodegradable, and usually don't reach into the groundwater. Fertilizers are more of a threat to surface water due to runoff.

Forestry management is another area of concern. As our population grows, the demand for lumber and wood products has grown exponentially. The increased urbanization has claimed once vast tracts of forests, replacing them with concrete paving and closely packed structures.

This same drive to urbanization also affects our soil. Arable farmland is shrinking as the press to develop home and commercial sites increases. Of the approximately 15 billion hectares of dry land on the Earth, only 2 billion are suitable for agriculture. If the same land is used year after year, there is a definite danger of soil exhaustion as vital nutrients are depleted. Grazing lands for our cattle and other domesticated animals are also shrinking, and as a consequence, many of the remaining areas are being overgrazed. The danger of overgrazing lays in the non-availability of sufficient pasturage for the animals and the loss of top cover for the soil, which is then left vulnerable to erosion.

**Air Pollution**

**Pollution**: the chemical or physical changes that are natural or man-made (anthropomorphic), that result in quality degradation.

**Air Pollution & History**

1273 - London, England: King Edward recognizes an air pollution problem. He issues a proclamation banning the burning of sea coal. Sea coal was a particularly poor grade of coal that gave off voluminous amounts of smoke.

1873 - London, England: 700 people die of the effects of the thick smoke caused by industrial activities.

1911 - The term "Smog" is coined and used in periodicals of the time.

1952 - The smog problem reaches a peak. 1000 people die of its effects. Parliament responds by passing the "Clean Air Act."

1948 - In the United States, 20 people die in Nora, Pennsylvania, due to air pollution.

1970 - The U.S. Congress passes the "Clean Air Act." The act has subsequently been revised and updated, but the economic costs of implementing the act have been enormous.

## Types of Pollutants

**Primary Pollutants**: chemicals released into the air in a harmful form.
Example: Smokestack exhaust is harmful if not properly scrubbed before release.

**Secondary Pollutants**: chemicals that are not harmful as released, but are modified in the atmosphere and become harmful.
Example: Exhaust emissions of Nitrous Oxide from automobiles. When exposed to sunlight they become harmful.

**Fugitive Emissions**: emissions released by mistake.
Example: Holes in automobile tail pipes, mufflers, catalytic converters, and failed industrial scrubber units.

## Sources of Outdoor Pollutants

Although there are natural pollutants, the primary sources of outdoor air pollution are pollutants created by humans. Transportation sources alone account for nearly 50% of all outdoor air pollution.

The U.S. **Environmental Protection Agency** (EPA) recognizes seven substances as being harmful to the environment.

**Sulfur Compounds**: Some are naturally released. The primary natural sources are volcanoes. The man-made sources are from the combustion of coal and oil products. When released, the sulfur combines with the atmosphere and changes to an oxidizer, Sulfur Dioxide ($SO_2$). Further chemical modifications transform this into Sulfate, and when combined with water, it changes to Sulfuric Acid ($SO_4$). This acid precipitation is popularly known as Acid Rain. The solution to the problem is to use low sulfur content coal. Since the EPA mandated its use, there has been an 11% drop in the amount of acid rain.

**Nitrogen Compounds**: 5% of all Nitrogen compounds occur naturally via bacterial decay, rice paddies, wetlands, and swamp decay. 95% of the problem is caused by man-made, high temperature combustion. Auto exhaust is the primary offender. The nitrogen compounds oxidize and combine with $H_2O$ (water) to form into $SO_4$ (acid rain).

**Carbon Oxides**: Carbon Monoxide (CO) is very deadly because it replaces oxygen molecules in the blood, causing Carbon Monoxide poisoning. Carbon Monoxide is produced by combustion of oil and gas, with auto exhaust as the primary offender.

**Particulates**: These are Aerosols: small bits of solids suspended in $H_2O$ (water) drops. Aerosols block solar radiation and make the air look dirty. The source of the particulate (dust, soot, ash) can be either man-made or natural. 40% comes from industrial processes. 17% comes from vehicle emissions. 30% comes from natural sources: salt spray, dust, and volcanoes. 13% comes from other miscellaneous sources. The most dangerous of the particulates are <2.5 microns in size. These microscopic sized particles cause lung problems. Some sources are asbestos, cigarette smoke, and coal dust. Example: Long exposure to coal dust can cause "Black Lung" disease. The sulfur in the dust particulates literally burns holes in the lungs.

**Metals**: Metal substances are the byproducts of a variety of manufacturing processes. Some sources are sandblasting, leaded gasoline, and leaded paint. Example: Since the EPA banned the use of lead in gasoline and paint there has been a drop in the lead accumulation problem observed in the environment. The major problem with metals is that almost all of them are carcinogenic, and because the metals accumulate in the fatty tissues of the body, they also cause long-term problems with nervous system disruption. Example: The haberdashery (hat maker) trade of the past had a high rate of craftsmen going insane. The reason was that the trade used to use a large amount of mercury in the crafting process. The fumes from the mercury were breathed in by the workers, and eventually caused severe nervous system disruption. Thus comes the old phrase, "As Mad as a Hatter".

**Photochemical Oxidants**: The oxidation process can change normally harmless chemicals into more dangerous substances. Example: Oxygen ($O_2$), when exposed to ultra-violet radiation (UV), becomes Ozone ($O_3$). Although Ozone is necessary in the upper atmosphere, when the human body is exposed to it, it causes lung damage. Example: Nitrogen compounds under certain circumstances can transform into $NO_3$. This compound is a major component of smog that irritates the soft tissues and mucous membranes.

**Volatile Organic Compounds (VOCs)**: These are organic chemical gases that occur both naturally and through man-made processes.

**Natural sources**: Plants, wetlands, rice paddies, and ruminant animals release Methane. Conifer trees release Terpene. Terpene forms a haze in the atmosphere. It is what forms the haze over the Great Smoky Mountains and the Azure Mountains.

**Man-made Sources**: Any synthetic organic chemical (such as Benzene, Toluene, and Formaldehyde). All hydrocarbons emissions such as gasoline fumes.

## Controlling Air Pollution Outdoors

**Cyclone Collectors**: Utilizes an electrostatic precipitator. Small particulates are electronically attracted to a charged filter screen and destroyed.

**Scrubber**: Used to control $SO_2$ (Sulfur Dioxide) emissions. Limestone is crushed and mixed into a slurry. The slurry mixture chemically removes 95% of the $SO_2$ emissions. Use of scrubber units is a relatively inexpensive procedure. However, the United States is the only country to consistently use this method. In 1987, only 40% of German industries used this method, compared to 95% of U.S. industries.

**Afterburners** and **Converters**: Nitrogen compounds in auto exhaust can be removed by the use of afterburners. Afterburners rely on another stage in the combustion process to rechannel the exhaust back to the combustion source and re-burn it. Catalytic converters use platinum to act as a catalyst to remove $CO_2$ and CO compounds.

## Indoor Air Pollutants

Indoor air pollution is on the rise due to urban lifestyles. New construction techniques cause better sealing of homes and buildings, trapping air inside of buildings. However, if the air filtration system isn't good or properly maintained, then the occupants end up re-breathing polluted air.

**Carbon Monoxide**: Produced by gas or oil burning stoves, water heaters, furnaces, fireplaces, etc., the gas is colorless, odorless, and deadly. Preventive safety measures include ensuring proper ventilation and the use of Carbon Monoxide (CO) detectors.

**PAH (Polycyclic Aromatic Hydrocarbons)**: This gas is given off by new carpeting, flooring, fabrics, and padding made from polycarbonates. Long-term exposure to PAH causes eye tearing and aggravates asthma and bronchitis.

**Radon**: A naturally occurring radioactive gas. The source of the emission is the natural decay of radioactive elements from granite, uranium, and other metamorphic rocks that contain radioactive elements. As the rock material weathers or breaks down, the gas seeps into the foundations of buildings and homes. The gas can then be re-circulated through the air system. The gas is a carcinogenic and, in areas where this type of material forms the bedrock, the use of Radon detectors is advised.

**Formaldehyde**: This chemical is used extensively in the processing of fabric. The fumes from trace amounts left on the fabric cause teary eyes and skin rashes.

**Asbestos**: A naturally occurring mineral, asbestos was once extensively used as a fire retardant in homes, buildings, schools, and as lining for automobile brake shoes. In its natural state as a solid, it is harmless. However, if the fibers become airborne, they can collect in human lungs and eventually cause asbestosis. Long term exposure to airborne particulate asbestos fibers can be deadly. In the short term, it can cause severe respiratory problems. Its use is now banned by the EPA, and the rip out and replacement process is very time consuming and expensive.

**Tobacco Smoke**: The EPA has identified this as a major carcinogen. The EPA has also identified second hand tobacco smoke as a problem. The effects of tobacco smoke are lung cancer, emphysema, heart disease, asthma, bronchitis, and Brittle Lung Disease (lungs lose their elastic capacity). EPA data is the major source of material cited in the enactment of antismoking legislation.

### SKILL 1.3    Describe strategies for dealing with environmental problems.

**Addressing the Issues**

Our increasing population, urbanization, and dependence on technology are the key factors that drive the rapid consumption of our resources. How long our natural resources will last depends on future demand and willingness on the part of governments to efficiently manage their energy needs and resources.

Likewise, industry must become more deeply involved by modifying their existing, or developing new, techniques and procedures to effectively utilize our natural materials.

Unfortunately, natural resources are not evenly distributed throughout the Earth, and political considerations have, to date, hampered cooperation of conservation efforts and development of alternative energy sources on a global scale.

As grim as the projected shortfalls may seem, there is some hope. There is a growing awareness of the problems we face and although not usually coordinated on a global scale, some countries are taking steps to address the issues.

Better agricultural techniques to prevent soil depletion, reclamation of waterways, banning the use of chemicals damaging to the atmosphere, recycling plastics and metals, and seeking alternative energy sources are all examples of ongoing initiatives to ensure resources for future generations.

## Controlling Air Pollution Indoors

The vast majority of problems caused to humankind by indoor air pollutants are directly attributable to bad ventilation. Filter systems are not cleaned on a regular basis or replaced frequently enough. If improper ventilation systems are installed they may restrict airflow or have insufficient exchange capability.

**Sick Building Syndrome**: Sick building syndrome is the name given to the effects from pollutants being re-breathed. New construction techniques more effectively seal buildings, trapping the pollutants inside, and can cause cold-like symptoms of watery eyes, fatigue, and respiratory problems. Trapped pollutants can become deadly.

**Legionnaires Disease**: Named after the incident in which many attendees at an American Legion convention died. Mold in the air conditioning system caused the death of the attendees. Areas of water damage can promote the growth of black mold (Staphybaucus bacteria) which causes bleeding in the human lung.

Good ventilation and properly maintained ventilation systems are the key factors in controlling indoor air pollution.

## Solutions to the Problems

Restoration of the Kissimee River & Everglades Wetlands: The Army Corps of Engineers (ACE) built a $30 million dollar canal project through the wetlands. Finished in 1971, the project has caused massive problems for the ecosystem. ACE is removing the canal and letting the land flood naturally. However, this is a slow and expensive process. The estimated cost is $600 million to acquire the 88,000 acres of land that had been developed.

The EPA is cracking down on the Orlando tourist complex. Addressing the problem of dumping of untreated wastewater, the city of Orlando was forced by the EPA to build a $200 million dollar sewage treatment plant. The plant treats and recycles 25-50 million gallons of wastewater per year. This treated water is used for irrigation.

Southern Florida instituted a Water Management District (WMD). The WMD determines water policy and monitors the use of water resources. Its policies are directed at preventing pollution, over use, etc.

## Treating Contaminated Groundwater

Some treatments are available to reclaim polluted groundwater.

- Pump the water out, haul it away, and treat it. However, this surmises that the water taken out will be replaced naturally. Of course this is self-defeating if the source of the pollution isn't eliminated.

- Biomediation: Use of contaminant eating enzymes.

- Using plastic barriers under landfills to prevent leeching.

- Treatment Beds: The concept is to dig down and place a silt-sand layer between the contaminated area and clear water. The layer filters the water and is now being used to protect many cities' water supplies. However, this is a very expensive process and is generally only employed by affluent municipalities.

## Sewage Treatment Sequence

lst Stage: Primary treatment. Skim off large objects by using screens and settle suspended solids. However, the waste is still rich in pathogens.

2nd Stage: Secondary treatment. Fine material in suspension is removed by trickling it over stones or activated sludge. Mixing the wastewater in a tank with active "friendly" bacteria forms activated sludge. The bacteria reduce the material by eating it.

3rd Stage: Chlorinate the water. Chlorination kills the bacteria and pathogens. However, non-biodegradable compounds are still in the water. Example: nitrogen and phosphorus from fertilizers, and oil and other hydrocarbons.

4th Stage: Tertiary treatment. Removes or binds up the heavy metals. This deals with fertilizer and hydrocarbon contamination. This is a very expensive stage and many communities do not use a tertiary treatment

**COMPETENCY 2.0**     Understand the nature of scientific inquiry, the role of observation and experimentation in science, and the relationships between earth science, technology, and other fields of knowledge.

**SKILL 2.1**     Identify the processing by which new scientific knowledge and hypotheses are generated and rejected.

Modern science began around the late 16th century with a new way of thinking about the world. Few scientists will disagree with Carl Sagan's assertion that "science is a way of thinking much more than it is a body of knowledge" (Broca's Brain, 1979). Thus, science is a process of inquiry and investigation. It is a way of thinking and acting, not just a body of knowledge to be acquired by memorizing facts and principles. This way of thinking, the scientific method, is based on the idea that scientists begin their investigations with observations. From these observations they develop a hypothesis, which is extended in the form of a prediction, and challenge the hypothesis through experimentation that results in further observations. Science has progressed in its understanding of nature through careful observation, a lively imagination, and increasingly sophisticated instrumentation. Science is distinguished from other fields of study in that it provides guidelines or methods for conducting research, and the research findings must be reproducible by other scientists for those findings to be valid. It is important to recognize that scientific practice is not always this systematic. Discoveries have been made that are serendipitous and others have not originated with the observation of data. Einstein's theory of relativity started not with the observation of data but with a kind of intellectual puzzle.

The Scientific Method is just a logical set of steps that a scientist goes through to solve a problem. There are as many different scientific methods as there are scientists experimenting. However, there is a pattern to their work.

**Observations**

Scientific questions result from observations of events in nature or events observed in the laboratory. An **observation** is not just a look at what happens. It also includes measurements and careful records of the event. Records could include photos, drawings, or written descriptions. The observations and data collection lead to a question. Having arrived at a question, a scientist usually researches the scientific literature to see what is known about the question. Maybe the question has already been answered. The scientist then may want to test the answer found in the literature. Or, maybe the research will lead to a new question.

Sometimes the same observations are made over and over again and are always the same. For example, you can observe that daylight lasts longer in summer than in winter. This observation never varies. Such observations are called **laws** of nature.

### Hypothesis

If the question has not been answered, the scientist may prepare for an experiment by making a hypothesis. A **hypothesis** is a statement of a possible answer to the question. It is a tentative explanation for a set of facts and can be tested by experiments. Although hypotheses are usually based on observations, they may also be based on a sudden idea or intuition.

### Experiment

An **experiment** tests the hypothesis to determine whether it may be a correct answer to the question or a solution to the problem. Some experiments may test the effect of one thing on another under controlled conditions. Such experiments have two variables. The experimenter controls one variable, called the **independent variable**. The other variable, the **dependent variable,** is the change caused by altering the independent variable.

### Conclusion

Finally, a scientist must draw conclusions from the experiment. A conclusion must address the hypothesis on which the experiment was based. The conclusion states whether or not the data supports the hypothesis. If it does not, the conclusion should state what the experiment showed. If the hypothesis is not supported, the scientist uses the observations from the experiment to make a new or revised hypothesis. Then, new experiments are planned.

### Theory

When a hypothesis survives many experimental tests to determine its validity, the hypothesis may evolve into a **theory**. A theory explains a body of facts and laws that are based on the facts. A theory also reliably predicts the outcome of related events in nature. For example, the law of conservation of matter and many other experimental observations led to a theory proposed early in the 19th century. This theory explained the conservation law by proposing that all matter is made up of atoms, which are never created nor destroyed in chemical reactions, only rearranged. This atomic theory also successfully predicted the behavior of matter in chemical reactions that had not been studied at the time. As a result, the atomic theory has stood for 200 years with only minor modifications.

A theory also serves as a scientific **model**. A model can be a physical model made of wood or plastic, a computer program that simulates events in nature, or simply a mental picture of an idea. A model illustrates a theory and explains nature. Outside of science, the word theory is often used to describe someone's unproven notion about something. In science, theory means much more. It is a thoroughly tested explanation of things and events observed in nature. A theory can never be proven true, but it can be proven untrue. All it takes to prove a theory untrue is to show an exception to the theory.

The test of the hypothesis may be observations of phenomena or a model may be built to examine its behavior under certain circumstances. Theories provide a framework to explain the known information of the time, but are subject to constant evaluation and updating. There is always the possibility that new evidence will conflict with a current theory.

Some examples of theories that have been rejected because they are now better explained by current knowledge:

- Theory of Spontaneous Generation
- Inheritance of Acquired Characteristics
- The Blending Hypothesis

Some examples of theories that were initially rejected because they fell outside of the accepted knowledge of the time, but are well accepted today due to increased knowledge and data include:

- The sun-centered solar system
- Warm-bloodedness in dinosaurs
- The germ-theory of disease
- Continental drift

**Law**

A Law is the highest level that science can achieve. Law is defined as a statement of an order or relation of phenomena that, so far as is known, is invariable under the given conditions. Everything we observe in the universe operates according to known natural laws.

Some well know and accepted natural laws of science are:

- The First Law of Thermodynamics
- The Law of Biogenesis
- The Law of Gravity

## SKILL 2.2 Understand ethical issues related to Earth science (e.g., accurately reporting experimental results).

To understand scientific ethics, we need to have a clear understanding of ethics. Ethics is defined as a system of public, general rules for guiding human conduct (Gert, 1988). The rules are general in that they are supposed to all people at all times and they are public in that they are not secret codes or practices.

Scientists are expected to show good conduct in their scientific pursuits. Conduct here refers to all aspects of scientific activity including experimentation, testing, education, data evaluation, data analysis, data storing, peer review, government funding, the staff, etc.

The following are some of the guiding principles of scientific ethics:

1. Scientific Honesty: not to fraud, fabricate, or misinterpret data for personal gain
2. Caution: to avoid errors and sloppiness in all scientific experimentation
3. Credit: give credit where credit is due and not to copy
4. Responsibility: only to report reliable information to public and not to mislead in the name of science
5. Freedom: freedom to criticize old ideas, question new research, and freedom to research

To discuss scientific ethics, we can look at natural phenomena, such as rain. Rain in the normal sense is extremely useful to us and it is absolutely important in the water cycle. When rain is polluted with acid, it becomes acid rain. Here lies the ethical issue of releasing all these pollutants into the atmosphere. Should the scientists communicate the whole truth about acid rain or withhold some information because it may alarm the public. There are many issues like this. Scientists are expected to be honest and forthright with the public.

## SKILL 2.3 Evaluate the appropriateness of a specified experimental design to test a given hypothesis in Earth science.

Only certain types of questions can truly be answered by science because the scientific method relies on observable phenomenon. That is, only hypotheses that can be tested are valid. Often this means that we can control the variables in a system to an extent that allows us to truly determine their effects. If we don't have full control over the variables, for instance, in environmental biology, we can study several different naturally occurring systems in which the desired variable is different.

The scientific method is particularly useful for determining 'cause and effect' type relationships. Thus appropriate hypotheses are often of this nature. The hypothesis is simply a prediction about a certain behavior that occurs in a system. Then variables are changed to determine whether the hypothesis is correct. For instance, let's consider several identical potted African violets and suppose we have lights of different color, fertilizer, water, and a variety of common household items. Below are some possible questions, phrased as hypotheses, and a bit about why they are, or are not, valid.

1. African violets will grow taller in blue light than they will in red light.
    *This hypothesis is valid because it could be tested by growing one violet in blue light and another in red. The results are easily observed by measuring the height of the violets.*

2. Invisible microbes cause the leaves of African violets to turn yellow.
    *This hypothesis is not valid because we cannot know whether a given violet is infected with the microbe. This hypothesis could be tested if we had appropriate technology to detect the presence of the microbe.*

3. Lack of water will stop the growth of African violets.
    *This hypothesis is also valid because it could be tested by denying water to one violet while continuing to water another. The hypothesis may need to be refined to more specifically define how growth will be measured, but presumably this could be done.*

4. African violets will not grow well in swamps.
    *This hypothesis is not valid in our specific situation because we have only potted plants. It could be tested by actually attempting to grow African violets in a swamp, but that is not within this scenario.*

**SKILL 2.4  Understand the role of communication among scientists in promoting scientific progress.**

Observations, however general they may seem, lead scientists to create a viable question and an educated guess (hypothesis) about what to expect. While scientists often have laboratories set up to study a specific thing, it is likely that along the way they will find an unexpected result. It is always important to be open-minded and to look at all of the information. An open-minded approach to science provides room for more questioning, and, hence, more learning. A central concept in science is that all evidence is empirical. This means that all evidence must be observed by the five senses. The phenomenon must be both observable and measurable, with reproducible results.

The question stage of scientific inquiry involves repetition. By repeating the experiment you can discover whether or not you have reproducibility. If results are reproducible, the hypothesis is valid. If the results are not reproducible, one has more questions to ask. It is also important to recognize that one experiment is often a stepping-stone for another. It is possible that data will be re-tested (by the same scientist or by another), and that a different conclusion may be found. In this way, scientific competition acts as a system of checks and balances.

Scientific theory and experimentation must be repeatable. It is also possible to be disproved and is capable of change. Science depends on communication, agreement, and disagreement among scientists. It is composed of theories, laws, and hypotheses.

The goal of the human genome project is to map and sequence the three billion nucleotides in the human genome, and to identify all of the genes on it. The project was launched in 1986 and an outline of the genome was finished in 2000 through international collaboration. In May 2006, the sequence of the last chromosome was published. While the map and sequencing are complete, scientists are still studying the functions of all the genes and their regulation. Scientists have successfully decoded the genomes of other mammals as well.

It is important to realize that many of the most complex scientific questions have been answered in a collaborative form. The human genome project is a great example of research conducted and shared by multiple countries world wide. It is also interesting to note that because of differing cultural beliefs and/or legislation, some countries may be more likely to allow certain areas of research that other cultures may be unlikely to examine.

**SKILL 2.5**      **Identify the similarities and differences between science and technology (e.g., science as investigating the natural world, technology as solving human adaptation problems).**

Science and technology, while distinct concepts, are closely related. Science attempts to investigate and explain the natural world, while technology attempts to solve human adaptation problems. Technology often results from the application of scientific discoveries, and advances in technology can increase the impact of scientific discoveries. For example, Watson and Crick used science to discover the structure of DNA and their discovery led to many biotechnological advances in the manipulation of DNA. These technological advances greatly influenced the medical and pharmaceutical fields. The success of Watson and Crick's experiments, however, was dependent on the technology available. Without the necessary technology, the experiments would have failed.

The combination of biology and technology has improved the human standard of living in many ways. However, the negative impact of increasing human life expectancy and population on the environment is problematic. In addition, advances in biotechnology (e.g. genetic engineering, cloning) produce ethical dilemmas that society must consider.

**Sciences are Interconnected**

Earth Science and Biology

Over the course of Earth's history, living things have been greatly affected by Earth processes. Volcanic eruptions, plate tectonics, and climate change have affected whether living things have survived or how they have had to adapt in order to survive. At least four of the major mass extinctions have been caused by a climatic change triggered by some Earth event, interconnecting geology, meteorology, and biology. The working explanation for both the Cretaceous-Tertiary extinction and the Permian-Triassic extinction include a large asteroid or meteorite impact. The majority of species prior to the impact were killed, and the remaining species had to adapt to the new environment in order to survive.

Earth processes have also affected how humans live. The most fertile land for farming is at the base of volcanoes. We have developed technologies to irrigate farmland, built "safe" buildings in earthquake prone areas, and prevented low - lying areas from flooding. Currently, humans' impact on the environment and on the Earth's temperature is being widely felt. All in all, biology has been greatly affected by Earth processes.

Earth Science and Chemistry

Earth science and chemistry are tightly woven. In geology, the chemical composition of the rocks and the temperature and pressure at which crystals form are an obvious connection between chemistry and Earth science. In addition, chemistry and oceanography are connected inherently. The salinity of Earth's oceans is affected by the temperature at which water freezes, the density of water, and the solubility of certain chemical compounds. Chemistry and meteorology are connected through the chemical makeup of the atmosphere and the effects that human released chemicals have on the atmosphere (i.e. CFC's effect on the ozone layer and carbon dioxide's role in climate change).

Earth Science and Physics

Earthquakes, plate tectonics, and meteorology are all related to physics. Fault production and earthquakes are caused by large-scale plate tectonics forming small-scale zones of weakness in the crust. The pressure builds up along fault lines due to increases in fault stress. At some point in time, the stress on the fault line will exceed the static frictional force of the fault line, and seismic waves will be released. The frictional force and the dynamics of the Earth's motion during earthquakes are all related to physics.

Meteorology and physics are closely related. Changes in atmospheric pressure cause winds, updrafts and storms. These pressure changes are caused by changes in temperature. Warm, moist air rises because it is less dense than the air surrounding it. As air rises and cools, it condenses, forming cloud systems. When meteorologists predict the weather, the physics of the interaction between volume, humidity, temperature, and pressure are studied in detail.

## COMPETENCY 3.0 Understand the processes of gathering, organizing, reporting, and interpreting scientific data in the context of Earth science investigations.

### SKILL 3.1 Evaluate the appropriateness of a given method or procedure for collecting data for a specified purpose.

Measurements may be taken in different ways. There is an appropriate measuring device for each aspect of science. A graduated cylinder is used to measure volume. A balance is used to measure mass. A microscope is used to view microscopic objects. A centrifuge is used to separate two or more parts in a liquid sample. The list goes on, but you get the point. For each variable, there is an appropriate way to measure it. The internet and teaching guides are virtually unlimited resources for laboratory ideas. You should be imparting on the students the importance of the method with which they conduct the study, the resource they use to do so, the concept of double checking their work, and the use of appropriate units.

### SKILL 3.2 Organize and report given experimental data using appropriate and effective graphic representation (e.g., graphs, tables, diagrams).

The type of graphic representation used to display observations depends on the data that is collected. **Line graphs** are used to compare different sets of related data or to predict data that has not yet been measured. An example of a line graph would be comparing the rate of activity of different enzymes at varying temperatures. A **bar graph** or **histogram** is used to compare different items and to make comparisons based on this data. An example of a bar graph would be comparing the ages of children in a classroom. A **pie chart** is useful when organizing data as part of a whole. A good use for a pie chart would be displaying the percentage of time that students spend on various after school activities.

### SKILL 3.3 Identify the procedures and criteria for formally reporting experimental procedures and data to the scientific community.

Most scientific studies are ultimately presented in technical journals (some well known examples include *Nature*, *Science*, and the *Journal of the American Medical Association*). Scientists prepare manuscripts detailing the conditions of their experiments and the results they obtained. They will typically also include their interpretation of those results and their impact on current theories in the field. These manuscripts are not unlike lab reports, though they are considerably more polished. Manuscripts are then submitted to appropriate technical journals.

All reputable scientific journals use peer review to assess the quality of research submitted for publication. Peer review is the process by which scientific results produced by one group are subjected to the analysis of other experts in the field. Reviewers of scientific work are typically experts in the field, but it is important that they be objective in their evaluations. Peer review is typically done anonymously so that the identity of the review remains unknown by the scientists submitting work for review. The goal of peer review is to "weed out" any science not performed by appropriate standards. Reviewers will determine whether proper controls were in place, enough replicates were performed, and that the experiments clearly address the presented hypothesis. The reviewer will scrutinize the interpretations and how they fit into what is already known in the field. Often reviewers will suggest that additional experiments be done to further corroborate presented conclusions. If the reviewers are satisfied with the quality of the work, it will be published and made available to the entire scientific community.

**SKILL 3.4    Identify relationships between factors (e.g. inverse, direct, linear) as indicated by experimental data.**

A direct relationship (also called a positive relationship) is observed when both variables increase (or decrease) at once. The key here is that they both go in the same direction concurrently. An inverse relationship (also called a negative relationship) is present when one variable increases (decreases) while the other variable decreases (increases). In this relationship, the variables are moving in opposite directions.

The individual data points on the graph of a linear relationship cluster around a line of best fit. In other words, a relationship is linear if we can sketch a straight line that roughly fits the data points. Consider the following examples of linear and non-linear relationships.

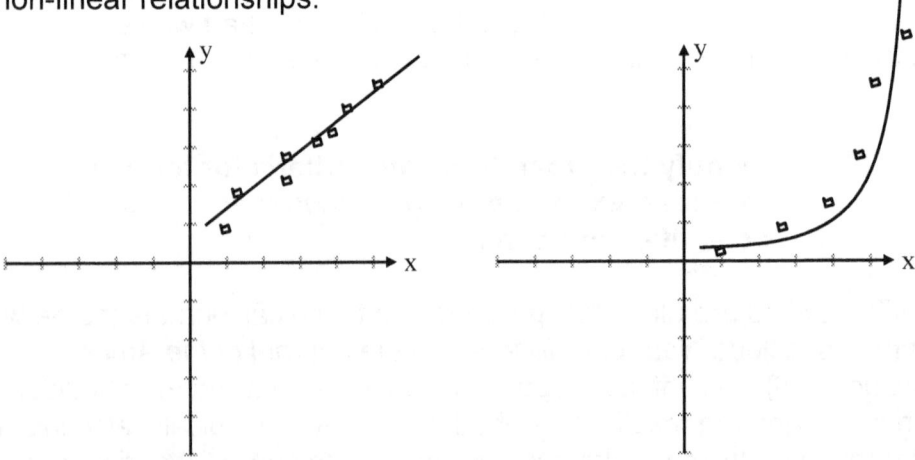

Linear Relationship          Non-Linear Relationship

Note that the non-linear relationship, an exponential relationship in this case, appears linear in parts of the curve. In addition, contrast the preceding graphs to the graph below which is of a data set that shows NO relationship between variables.

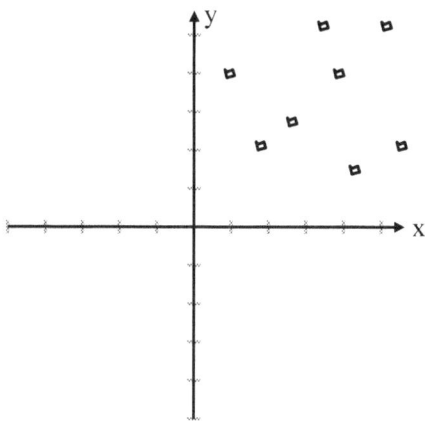

Extrapolation is the process of estimating data points outside of a known set of data points. When extrapolating data of a linear relationship we extend the line of best fit beyond the known values. The extension of the line represents the estimated data points. Extrapolating data is only appropriate if we are relatively certain that the relationship is indeed linear. For example, the death rate of an emerging disease may increase rapidly at first and level off as time goes on. Thus, extrapolating the death rate as if it were linear would yield inappropriately high values at later times. Similarly, extrapolating certain data in a strictly linear fashion, with no restrictions, may yield obviously inappropriate results. For instance, if the number of plant species in a forest were decreasing with time in a linear fashion, extrapolating the data set to infinity would eventually yield a negative number of species, which is clearly unreasonable.

**COMPETENCY 4.0**  Understand how to create, use, and interpret physical and mathematical models (e.g. maps, charts, graphs, diagrams) commonly used in Earth science.

**SKILL 4.1**  Describe alternative models for conveying given information from Earth science.

Generally speaking, a model is a proxy or entity used to represent some other entity. A scientific model is a simplified or idealized description or construct that represents a particular system, situation, process, or phenomenon. Models generally consist of a set of variables with a set of logical and/or quantitative relationships between them. Models are constructed to enable reasoning about scientific processes within an idealized framework. Models are a very important part of scientific inquiry, as they are more accessible to study than is the system modeled, and changes in the structure of a model are easier to implement, isolate, understand, and communicate to others.

From a philosophical standpoint, there are four types of models: iconic, idealized, analogical, and phenomenological.

**Iconic**

An iconic model is a naturalistic replica or truthful mirror image of the target system. For example, an iconic model may be a scale model of the solar system, which is a downsized copy of the original system.

**Idealized**

An idealization is a deliberate simplification of a complicated matter in order to make the subject more easily understood. Most idealization models are one of two types. One type of idealization model strips away properties from a concrete object that are not relevant to the problem at hand, allowing the observer to focus on a limited set of properties in isolation. This method may also be referred to as Aristotelian abstraction, negligibility assumption, or method of isolation. The second type of idealization model involves deliberate distortion of representation, and is known Galilean idealization.

**Analogical**

Analogical models utilize one commonly known and easily understood system to represent a more complex or abstract system. These models are created based on relevant similarities between the two systems, such as shared properties or resemblance of relations between parts of two systems.

## Phenomenological

Phenomenological models represent observable properties of their targets without postulation of hidden mechanisms.

Models commonly found in the Earth Science classroom may include:

### Mental models

Ideas, such as an individual's personal relationship with the environment, that are used to guide behavior.

### Iconic models

Pictures, structures or drawings, such as a map of natural resource availability.

### Linguistic models

Verbal or written descriptions of information (such as the protocol for an Earth Science experiment), that can be used to improve classroom procedures.

### Physical object models

This type of model may include scale models, analog models, or prototypes.

### Mathematical or computer models

These models are systems of equations or logical expressions that predict the end products of chemical reactions or computer programs that simulate Earth Science processes, such as the water cycle. Mathematical models may also include graphs and charts.

**SKILL 4.2** **Describe methods by which given physical and graphic models are created.**

### Physical Models

A physical model provides a physical representation of an object, concept, or system. Physical models are created by first determining what aspects of a concept are to be represented, and at what scale. Qualitative and quantitative information related to these aspects must be collected and properly analyzed. Such analysis often entails the entering of collected information into a computer-modeling program that generates three-dimensional models from observation input.

A physical model of something large is usually smaller and relative size may be a very important concept in generating the model. For example, while a physical model of the Solar System is clearly smaller than the actual Solar System, planet sizes and distances will be scaled down relative to each other. In many cases of smaller scale models, size is only approximate or may even be intentionally distorted. A physical model of something very small usually provides a larger view of the object. The purpose of such models may be to see the structure of an object, for example, the wing of an insect that is not normally visible or properly visible to the human eye. For this reason, a consideration of detail is crucial when creating large-scale models.

A physical model may represent inner parts of an object or system that are concealed by an exterior casing. For example, physical models of human anatomy may be created to allow students to study the body layer by layer, moving from the exterior skin to the internal organs. When generating such models, the sequential layering of the body must be accurately noted and portrayed in the model. In general, the creation of physical models is critically based on the interrelationships of a system's or object's parts.

**Graphical Models**

Graphical models combine probability theory and graph theory to provide a natural tool for dealing with uncertainty and complexity, two major issues in applied mathematics and science. This type of model may provide the user with a visual representation of dependencies and correlations among random variables, or a computer generated and easily manipulated representation of a system to be studied. A computer aided design (CAD) system can be used to generate graphical models of two or three-dimensional objects. These programs are designed to allow variable input dimensions, enabling graphic models to represent moving parts, to portray various scenarios and to provide the option of interaction between operator and model. Quantitative data, such as geometry and dimension observations, are entered into such programs by the operator through standard input means including keyboards, graphic tablets, etc. Following data entry, models and drawings are plotted according to the CAD's particular analysis method and purpose.

**SKILL 4.3**        Classify different types of maps (e.g. topographic, geologic) used in Earth science and analyze the information conveyed by that type of map.

**Topographic Maps**

**Topographic maps** use fine lines drawn in ordered patterns to show the topography and elevation of the land.

**Topographic Map Symbols**

A series of special symbols and lines are used to display information about the shape and elevation of the landscape. There is a set of rules that determine how these symbols and lines are drawn on the map.

**Contour Interval**: Shows the amount of elevation between contour lines.

**Bench Mark**: Shows exact elevation. Often marked with a solid triangle.

**Contour Line**: Connects points that have the same elevation. Contour lines are closed loops, although all of the loop may not be visible on the map.

**Index Contours**: The heaviest contour lines, each marked with an elevation.

**Elevation**: A numerical indication of the contour line's elevation.

**Hachure Marks**: Short lines drawn inside a closed loop that indicate a depression. The marks point down slope.

**Gradient**: The relative spacing of the contouring lines indicates the gradient of the slope. The closer the lines, the steeper the gradient.

**Sea Level**: Elevation is measured as either above or below sea level.

**The Five General Rules of Contouring**

1. All contour lines either close or extend to the edge of the map.
2. Contour lines are closed around hills, basins, or depressions.
3. Contour lines never cross, although they are sometimes very close. The closer they are, the steeper the slope.
4. Contour lines appear on both sides of an area where the slope reverses direction.
5. Contour lines form V's that point upstream when they cross a stream, river, or valley.

The topographic map **legend** provides a great deal of information about the map, including the scale of the map, the agency that created the map, and the year it was created.

In the upper right hand corner of the map, another legend provides the geographic name of the area covered, latitude and longitude information, and the **minute series**- the relative coverage of the map. The larger the minute series number, the larger the area covered by the map. If a topographic map shows open bodies of water, the elevation markings in the water are given as depth soundings.

**Geologic Maps**

A **geologic map** is a special purpose map made to show subsurface geological features. In the United States, geologic maps are usually superimposed over a topographic map (and at times over other base maps) with the addition of a color mask and letter symbols to represent the kind of geologic unit, stratigraphic contour lines, fault lines, strike and dip symbols, and various additional symbols as indicated by the map key.

The most striking feature of a geologic map is its colors. Each color represents a different geologic unit. A geologic unit is a volume of a certain kind of rock of a given age range. Sandstone of one age might be colored bright orange, while sandstone of a different age might be colored pale brown. The capital letter represents the age of the geologic unit. Geologists have divided the history of the Earth into Eons, Eras, Periods, and Epochs. The most common division of time used in letter symbols on geologic maps is the Period. Most letter symbols begin with a capital letter representing one of the four Periods: J (Jurassic), K (Cretaceous), T (Tertiary), or Q (Quaternary). Sometimes the age of a rock unit will span more than one period. In that case, both capital letters are used. For example, QT would indicate that the rock unit began to form in Tertiary time and was completed in Quaternary time. The small letters indicate either the name of the unit, if it has one, or the type of rock, if the unit has no name.

The place where two different geologic units are found next to each other is called a contact, and that is represented by different kinds of lines on the geologic map. The two main types of contacts shown on most geologic maps are depositional contacts and faults. All geologic units are formed over, under, or beside other geologic units. When different geologic units have been moved next to one another after they were formed the contact is a fault contact, which is shown on the map by a thick line. Another kind of line shown on most geologic maps is a fold axis. In addition to being moved by faults, geologic units can also be bent and warped by the same forces into rounded wavelike shapes called folds. A line that follows the crest or trough of the fold is called the fold axis. This is marked on a geologic map with a line a little thicker than a depositional contact, but thinner than a fault.

All thicknesses of lines are also modified by being solid, dashed or dotted. Often contacts are obscured by soil, vegetation or human construction. Those places where the line is precisely located are shown as solid, but where it is uncertain it is dashed. The shorter the dash, the more uncertain the location. The lines on the map may also be modified by other symbols on the line (triangles, small tic marks, arrows, etc.) which give more information about the line. For example, faults with triangles on them show that the side with the triangles has been thrust up and over the side without the triangles. All the different symbols on the lines are explained in the map key.

A geologic map shows the distribution of rocks at the Earth's surface. However, bedrock is usually obscured so that only a small amount of outcrop is available for observation, study, or sampling. The geologist must then extrapolate the general distribution of rock types. Not only will the rock he can see help him, but so will changes in soil, vegetation and landscape, as well as various patterns detected from aerial photographs.

A geologic cross section map tells how the rocks are arranged underground. Each type of rock has its own symbol for shading and various structures have defined symbols. The geologic map and its cross section are used together to give a more complete picture.

**SKILL 4.4** **Interpret diagrams relating to earth science (e.g. crustal movements).**

Mountains are formed due to processes related to plate tectonics. When two plates collide, the continental crust is compressed and stressed, and is crumpled and pushed both upward and downward. The result is mountain ranges extending above the surface of the continent, as well as thicker continental crust below the mountains, known as mountain roots. Volcanic activity can also produce individual mountains or chains of mountains. Various types of mountains and causes of their formation are shown and described below.

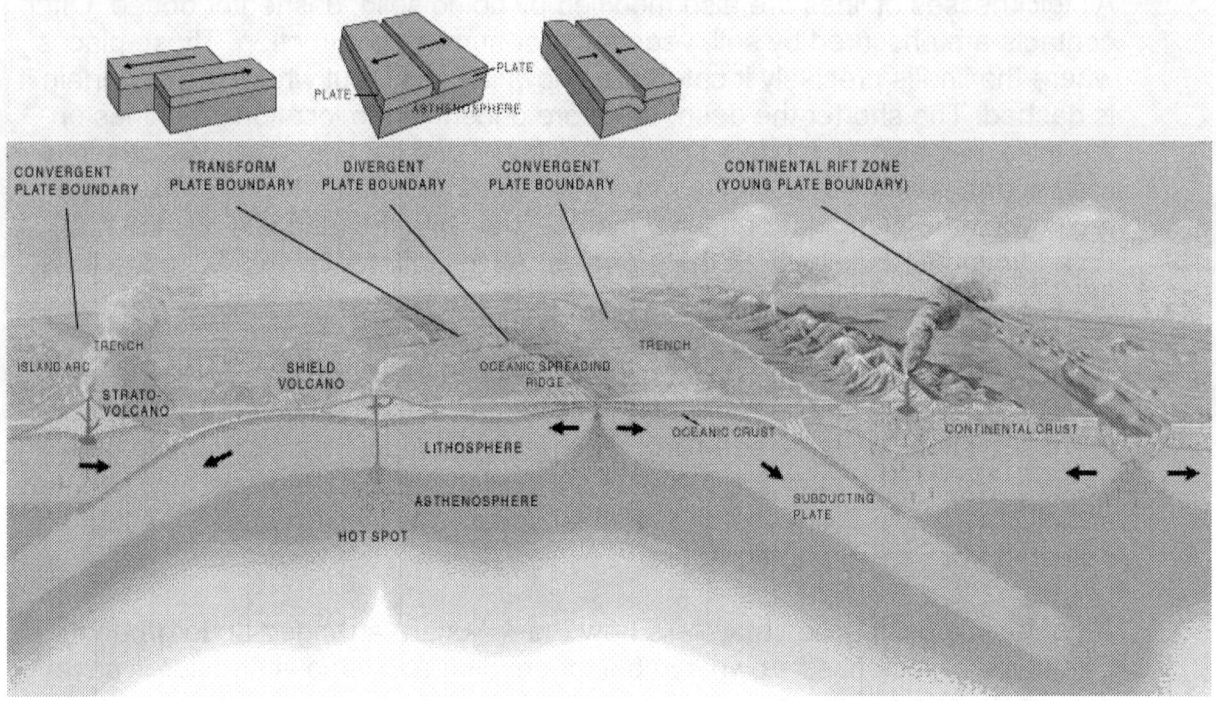

- **Collision of two continental plates** – The collision of two continental plates causes the continental crust at the contact point to become greatly deformed. Because the crust from both plates has no place to go or way to be destroyed, it can create very high mountain ranges such as the Himalayas.

- **Collision of a continental plate with an oceanic plate** – When an oceanic plate is subducted underneath a continental plate, two types of mountains can be formed. The continental crust may become stressed and deformed upward as above due to stresses and friction from the oceanic plate. In addition, the oceanic plate forms magma as it is subducted, and this magma may rise and create coastal volcanoes, either onshore or offshore. The so-called "Ring of Fire" of volcanoes surrounding the Pacific Ocean and associated coastal mountain ranges are of this nature.

- **Mid-ocean volcanic activity** – In locations known as "hot spots," magma rises through the crust and creates volcanic activity. Mountains formed by hot spots are usually located in the ocean, where the crust is thinner. Because the oceanic crust is always spreading, this intermittent volcanic activity can create a chain of islands, such as the Hawaiian Islands. When a hot spot is located under continental crust, magma may rise and cool before it reaches the surface, or it may create geothermal activity, such as at Yellowstone National Park.

## COMPETENCY 5.0     Understand proper and safe use of equipment and materials used in earth science investigations.

### SKILL 5.1     Identify the operating principles for various instruments.

Light microscopes are commonly used in high school laboratory experiments. The total magnification is determined by multiplying the ocular (usually 10X) and the objective (usually 10X on low, 40X on high). Several procedures should be followed to properly care for this equipment.

- Clean all lenses with lens paper only.
- Carry microscopes with two hands (one on the arm and one on the base).
- Always begin focusing on low power and then switch to high power.
- Store microscopes with the low power objective down.
- Always use a coverslip when viewing wet mount slides.
- Bring the objective down to its lowest position and then focus by moving upward to avoid breaking the slide or scratching the lens.

**Wet mount slides** should be made by placing a drop of water on the specimen and then putting a glass coverslip on top of the drop of water. Dropping the coverslip at a forty-five degree angle will help to avoid the presence of air bubbles on the slide.

**Chromatography** uses the principles of capillary to separate substances such as plant pigments. Molecules of a larger size will move slower up the paper, whereas smaller molecules will move more quickly, producing lines of pigment.

An **indicator** is any substance used to assist in the classification of another substance. An example of an indicator is litmus paper. Litmus paper is a way to measure whether a substance is acidic or basic. Blue litmus turns pink when an acid is placed on it and pink litmus turns blue when a base is placed on it. pH paper is a more accurate measure of pH, with the paper turning different colors depending on the pH value.

**Spectrophotometry** measures percent of light absorbed and transmitted by a pigment solution at different wavelengths.

**Centrifugation** involves spinning substances at a high speed. The more dense part of a solution will settle to the bottom of the test tube, while the lighter material will stay on top. Centrifugation is used to separate blood into blood cells and plasma, with the heavier blood cells settling to the bottom.

**SKILL 5.2**  Describe basic safety procedures in a laboratory or field situation ( e.g. wearing safety glasses).

Safety in the science classroom and laboratory is of paramount importance to the science educator. The following is a general summary of the types of safety equipment that should be made available within a given school system as well as general locations where the protective equipment or devices should be maintained and used. Please note that this is only a partial list and that your school system should be reviewed for unique and/or site-specific hazards at each facility.

*Safety glasses/goggles.* Safety glasses or goggles should be used during all activities where chemicals or particles may accidentally enter the eye. Common settings include middle and high school science labs, art studios, maintenance and shop areas. Be sure to review and consider the types of hazards present when selecting appropriate eye protection. Make sure you select equipment that is designed to provide protection for the risks under consideration.

*Gloves.* Many chemicals can either damage the skin via contact or may cause harm if absorbed through the skin. As a result, it is important to use protective gloves to form a barrier against these problems. Also, remember that different hazards require the use of different gloves. It is rare that one type of glove would be capable of providing adequate protection for all of the hazards that may be encountered. This is especially true with regard to the variety of chemistry that may be found in science laboratories. Common settings include middle and high school science labs, art studios, maintenance and shop areas.

*Lab coats and aprons.* Lab coats and aprons are designed to provide protection in the event of a spill. In addition, the use of these articles may prevent student or staff clothing from becoming contaminated and thereby limit the migration of chemical contamination should a spill occur. Common settings include middle and high school science labs, art studios, maintenance and shop areas.

*Eye wash units.* Eye wash units should be provided in all areas where chemical or physical hazards exist which may cause eye irritation or injury.

*Deluge showers/fire blankets.* Deluge showers and fire blankets should be maintained in all areas where chemical or fire hazards exist which may result in partial body exposure.

*Fire extinguishers.* Fire extinguishers should be maintained in accordance to state and local fire and building codes. Furthermore, each area should be reviewed individually to determine if additional fire protection is required. Be sure to verify that the correct size and type of extinguisher is provided in accordance to the materials commonly used or maintained in that space.

*Ventilation.* Additional mechanical ventilation, such as chemical fume hoods, should be provided in all areas where chemical fumes, vapors, or odors are commonly generated. Sufficient ventilation should be provided to prevent the build up of hazardous air contaminants and to minimize nuisance odors. This may be an issue in middle and high school art, science, and shop classrooms as well as facilities' maintenance shops.

*Respirators.* Respirators are not generally recommended for student use, but their use by maintenance staff may be required when conducting certain activities such as pesticide application or asbestos management.

*Hearing protection.* Certain lab and maintenance activities may necessitate the need to wear hearing protection. Each school system is encouraged to investigate all activities that generate loud or persistent noises for evaluation by a trained/certified industrial hygienist.

*Spill kits.* **Spills will occur!** Be prepared to address and respond to the problem. This approach will enhance your ability to contain a spill and minimize its long-term impact.

*Chemical storage cabinets.* The bulk storage of hazardous materials may necessitate the use of a re-enforced chemical storage cabinet. This method of storage is routinely used to manage flammables, corrosives, and oxidizers. Review local fire prevention guidance for storage recommendations.

**SKILL 5.3   Apply proper procedures for dealing with given accidents and injuries in the Earth science laboratory or in the field.**

All students and staff should be trained in first aid in the science classroom and laboratory. Please remember to **always report all accidents, however minor, to the lab instructor immediately.** In most situations **911** should immediately be called. Please refer to your school's specific safety plan for accidents in the classroom and laboratory. The classroom/laboratory should have a complete first-aid kit with supplies that are up-to-date and checked frequently for expiration.

Know the location and use of fire extinguishers, eye-wash stations, and safety showers in the lab.

Do not attempt to smother a fire in a beaker or flask with a fire extinguisher. The force of the stream of material from it will turn over the vessel and result in a bigger fire. Just place a watch glass or a wet towel over the container to cut off the supply of oxygen.

If your clothing is on fire, **do not run** because this only increases the burning. It is normally best to fall on the floor and roll over to smother the fire. If a student whose clothing is on fire panics and begins to run, attempt to get the student onto the floor and encourage him/her to roll over to smother the flame. If necessary, use the fire blanket or safety shower in the lab to smother the fire.

Students with long hair should put their hair in a bun or a pony-tail to avoid their hair catching fire.

Below are common accidents that everyone who uses the laboratory should be prepared for.

- **Burns (Chemical or Fire)** – Use deluge shower for 15 minutes.
- **Burns (Clothing on fire)** – Use safety shower *immediately*. Keep victim immersed 15 minutes to wash away both heat and chemicals. All burns should be examined by medical personnel.
- **Chemical spills** – Chemical spills on hands or arms should be washed immediately with soap and water. Washing hands should become an instinctive response to any chemical spilled on hands. Spills that cover clothing and other parts of the body should be drenched under the safety shower. If strong acids or bases are spilled on clothing, the clothing should be removed.
- If a large area is affected, remove clothing and immerse victim in the *safety shower*. If a small area is affected, remove article of clothing and use deluge shower for 15 minutes.
- **Eyes (chemical contamination)** – Hold the eye wide open and flush with water from the eye wash for about 15 minutes. Seek medical attention.
- **Ingestion of chemicals or poisoning** – See antidote chart on wall of lab for general first-aid directions. The victim should drink large amounts of water. All chemical poisonings should receive medical attention.

## SUBAREA II. ASTRONOMY

**COMPETENCY 6.0** Understand the physical characteristics and motions of the Earth as well as the evidence of and methods used to determine these characteristics and motions.

**SKILL 6.1** Describe historical methods used to study the characteristics and motions of the Earth (e.g., Foucault pendulum).

### Measuring distance

In 1400 BC, the Greek Herodotus wanted to survey land for the estimate of taxes. He had surveyors tie knots in ropes at set intervals. Then "rope stretchers" attached plumb bobs and used sighting instruments to form accurate measurements.

Eventually the Greeks abandoned the ropes, and used a triangulation method and built instruments in order to measure the angles. The Romans used 10 foot wooden rods with metal ends and laid them end to end in order to measure distances.

Edmund Gunter invented a survey chain in 1620 that was comprised of 100 iron links measuring 66 ft. long. This allowed him to measure more accurately uneven areas (such as a property line along a stream).

The first official survey of the east coast of the United States was done by Ferdinand Hassler. He standardized iron bars that were not as affected by temperature and humidity and clamped them together to be exactly 8 meters long. These would be laid down to measure distances. These rods were cumbersome and slow to use, and "invar" was created as a nickel steel hybrid. The invar was used to make metal tapes that did not have the expansion and contraction problems of steel.

In 1948, it was discovered that light could be used to measure distance. In the 1970's Doppler was used to measure distance. Finally, GPS and satellites are currently used with extreme accuracy. Through proper measurement, accurate maps were created and property lines were delineated.

### Measuring time

Prehistoric man used changes in seasons, stars, and day and night to measure time. This helped prehistoric man plan growing seasons and nomadic activity.

The sundial was one of the first formal technologies for measuring time. The sundial is a round disk that has a wedge on it. Around the circle are time markings. As the sun rises in the sky, the wedge casts shadows around the dial, measuring time against the markings.

The hourglass and the clepsydra (water clock) worked in similar ways. They both worked by allowing sand or water drip through a narrow opening to measure time.

Early clocks measured time by falling weights or springs. Then, pendulums were used to tick off time. Currently electronic motors and the vibration of quartz when a current is passed through it are used to measure the passing of time. Measuring time allowed civilizations to form calendars, plan religious ceremonies, and plan farming and hunting times of the year.

**Foucault's Pendulum**

Foucault's pendulum, named after the French physicist Léon Foucault, was created in 1851 to demonstrate the rotation of the Earth, and provided the first easily observable proof of this phenomenon. This apparatus consists of a tall pendulum that oscillates freely in a vertical plane, such as a bob attached to a cable or wire. Over time, the direction along which the pendulum swings changes due to the Earth's daily rotation. A Foucault pendulum located at the North Pole would demonstrate a fixed plane of oscillation with respect to the fixed stars, while the Earth rotates underneath it. This pendulum would require one sidereal day to complete a rotation. Relative to Earth, the plane of oscillation of the pendulum undergoes a full clockwise rotation during one day. A Foucault pendulum at the South Pole would experience a counter-clockwise rotation. At the Equator, a Foucault pendulum would demonstrate a fixed plane of oscillation, with a rotation of zero relative to Earth.

At other locations on the Earth's surface, the angular speed of rotation of the swinging plane is proportional to the sine of the geographical latitude. The general formula for alpha, the angle of rotation (measured in a counterclockwise direction) that the pendulum undergoes during one day at latitude φ, is:

$$\alpha = -2\pi \sin(\theta).$$

For example, a Foucault pendulum at a latitude of 30° rotates 180° in one day.

To solve a Foucault pendulum problem, the following coupled differential equations can also be used:

$$x = 2\Omega \sin\lambda \cdot y - (g/L) x$$

$$y = -2\Omega \sin\lambda \cdot x - (g/L) y$$

where

x,y = pendulum bob coordinates as seen by an observer on Earth
$\Omega$ = Earth's angular speed of revolution about its axis (rad/sec)
g = acceleration of gravity (m/sec$^2$)
L = length of the pendulum string (m)
$\lambda$ = geographic latitude

**SKILL 6.2** **Identify the physical characteristics of the Earth (e.g., diameter, tilt of axis, distance from the sun) and how they can be determined.**

| EARTH | |
|---|---|
| Average Distance from the Sun | 1.00 AU (1.495 x 10$^8$ km) |
| Density | 5.497 g/cm$^3$ |
| Diameter (equatorial) | 12,756 km |
| Mass | 5.976 x 10$^{24}$ kg |
| Atmosphere | 78% Nitrogen, 21% Oxygen |
| Axial Rotation Period | 24.00 hours |
| Axial Tilt | 23.5° |
| Surface Temp | -50°C to 50°C |
| Solar Revolution Period | 365.26 days |
| Moons | One, the Moon |

The density of planets correlates with their distance from the Sun. The inner planets (Mercury-Mars) are known as the terrestrial planets because they are rocky, and the outer planets (Jupiter and outward) are known as the icy or Jovian (gaslike) planets.

The Earth's inner core is mathematically hypothesized to be a solid iron and nickel core. The outer core, surrounding the inner core, is so hot that it is believed to be molten iron (liquid state). Combined, they are responsible for Earth's magnetism.

**SKILL 6.3**      **Describe the consequences of the Earth's motions (e.g., length of day, change in seasons, length of year).**

Seasonal change on Earth is caused by the orbit and axial tilt of the planet in relation to the Sun's Ecliptic: the rotational path of the Sun. These factors combine to vary the degree of insolation at a particular location and thereby change the seasons.

Equinox and Solstice

There are four key points on the Ecliptic. These are the Equinoxes and the Solstices.
- Winter Solstice (December 21) = Shortest day of the year in the northern hemisphere.
- Summer Solstice (June 21) = Longest day of the year in the northern hemisphere.
- Vernal Equinox (March 21) = Marks beginning of Spring.
- Autumnal Equinox (Sept 21) = Marks beginning of Autumn.

These dates will vary slightly in relation to leap years.

During the summer solstice, insolation is at a maximum in the northern hemisphere, and at a minimum in the southern hemisphere. Because of the tilt and curvature of the Earth, in order to get the sun directly overhead, you must be between 23.5°N Latitude and 23.5°S Latitude. This is between the Tropic of Cancer and the Tropic of Capricorn. Another result is that during the summer months in the northern hemisphere, the far northern latitudes receive 24 hours of daylight. This situation is reversed during the winter months, when it experiences 24 hours of darkness.

# TEACHER CERTIFICATION STUDY GUIDE

**COMPETENCY 7.0**     **Understand the structure, composition, and features of the Sun (including its production and transmission of energy) and the importance of the Sun to Earth processes.**

**SKILL 7.1**     Describe the methods by which the Sun's diameter, surface temperature, and chemical composition are determined.

## Sun's Diameter

When viewing the night sky, the human eye lacks depth perception. For this reason, many stars and planets appear to be of the same size and at the same distance from Earth. This phenomenon is known as the celestial sphere: a gigantic sphere of which Earth is the center. For example, the Sun is 400 times larger than the moon, but because it is also 400 times more distant from Earth, the Sun appears to be the same size as the full Moon.

The apparent sizes of and distances between objects are described with angular measurement, a system of measurement used by astronomers that treats the celestial sphere as a 360-degree circle, in which each degree is divided into 60 arc minutes and each arc minute is divided into 60 arc seconds. Using this system, the Sun and the Moon have angular diameters of about half a degree. Because scientists are able to accurately measure the distance between Earth and the Sun, it is possible to calculate the Sun's diameter according to the following formula:

$$\delta = 2 \arcsin[(1/2)(d/D)]$$

Where $\delta$ = angular diameter, and d and D are the diameter of and the distance to the object, respectively.

This formula demonstrates that even though the Moon and the Sun have the same angular size, the Sun has a radius of 696,000 km and the Moon has a radius of 1740 km.

EARTH SCIENCE

## Sun's Surface Temperature

The Sun's surface temperature can be determined by first measuring the amount of energy received on Earth per unit time over a unit area facing the Sun. Knowing the distance to the Sun, the total energy emitted by the Sun can then be determined. The total energy emitted per unit time per unit area is related to temperature, and can be used to determine the Sun's surface temperature.

Another method of estimating the Sun's surface temperature is to examine the Sun's light spectrum. The light emitted from the Sun is produced deep within the Sun's core. The Sun's surface is defined as the mean distance from the core at which the light bounces for the last time before emission. The Sun's surface demonstrates a light spectrum very similar to that of a blackbody, in which the fraction of light emitted at each wavelength is a function of the surface temperature alone. By measuring the spectrum of the Sun and fitting it to the spectrum of a blackbody, the Sun's surface temperature can be derived, and is known to be approximately 5880 Kelvin.

## Sun's Core Temperature

The Sun consists of gas-like particles. Without gravity, these particles would drift apart, but without an internal source of energy, these particles would compress to form a much smaller volume. The Sun must, therefore, possess both forces. The Sun's size is a balancing of these two forces, the gravity of the Sun that pulls the gas-like particles inward, and the energy output of the Sun (known to be produced by nuclear fusion) that pushes the gas-like particles outward.

By observing how other planets orbit the Sun, the Sun's mass can be determined because of the known relationship between mass and gravity. The Sun's diameter can be determined using angular measurement.

The following formula equates the pressure of a gas to the volume it takes up:

$$dP/dr = -\rho GM(r)/r^2$$

Where P is the pressure at radius r, rho is the density and G is the gravitational constant. M(r) is the mass interior to radius r, as this is the material that pulls down on a particle at radius r.

Pressure can then be calculated using:

$$P = nkT$$

Where n is the number density particles, k is the Stefan-Boltzmann constant and T is the temperature in degrees Kelvin.

This allows for the equation:

$T_c$ (Core Temperature) $\sim (G\, M_{sun} / R_{sun})(m_p/k)$

Where k is the Stefan-Boltzmann constant, $m_p$ is a proton mass and $M_{sun}$ and $R_{sun}$ are the solar mass and radius.

This gives $T_c \sim (6.67e\text{-}8 * 2e33/7e10)(1.67e\text{-}24/1.38e\text{-}16) \sim 2.3e7$ K

Therefore, by knowing mass of the Sun and the size of the Sun, one can determine how much energy the Sun is giving off, and relate this energy to the Sun's core temperature.

## SKILL 7.2 Identify the layers of the Sun.

The Sun is intensely hot. At the center, it has a 140,000-kilometer diameter core composed of hydrogen (92%) and helium (7.8%) that provide the fuel for the Sun's nuclear reaction (fusion). At approximately 15 million °C the core gives off a tremendous amount of energy. However, the density of the Sun precludes the direct release of all this energy into space. Instead, it is slowly absorbed and re-emitted by the various layers of the Sun. The first layer above the core is the very thick Radiative Layer. The energy produced in the core warms this layer to an average temperature of 3 million °C. On top of the radiative layer is the Convective Layer, where, as the name implies, energy is transferred via convection. This layer has an average temperature of 8,000 °C. The Sun's atmosphere comprises its visible layers. The atmosphere is made up of three layers.

The first layer is the Photosphere: the inner layer of the Sun's atmosphere that forms the Sun's visible surface. The photosphere is a very thin layer, only 400 kilometers deep and its average temperature is around 5,500 °C.
The photosphere's many small (1,000 kilometer in diameter), bright areas are referred to as Granules: the tops of rising columns of hot gas. The dark granules represent sinking columns of cooler gasses.

The second layer of the atmosphere is the Chromosphere: a 2,500-kilometer thick layer of turbulent gases. The temperature in the chromosphere increases to 100,000 degrees Celsius. Spicules: jets of heated gasses shoot upward from the chromosphere, reaching average proportions of 7,000 kilometers wide by 7,000 kilometers tall.

The third and final layer of the Sun's atmosphere is called the Corona: a very thin layer of gas that merges with outer space. The Corona's temperature is approximately 2 million degrees Celsius, and it is heated by twisted magnetic fields that carry energy up to it. These energy particles move very fast, generating high temperatures. Prominences: red, flaming jets of gas that rise from the corona, can travel as far as 1 million kilometers outward into space. Electrons are stripped off of atoms and they, and the reduced nuclei, fly off into space, creating the Solar Wind. This wind surrounds the Earth and even reaches to the farthest regions of the solar system.

**SKILL 7.3**      **Describe the Sun's source of energy (fusion reaction).**

All stars derive their energy through the thermonuclear fusion of light elements into heavy elements. The minimum temperature required for the fusion of hydrogen is 5 million degrees. Elements with more protons in their nuclei require higher temperatures. For instance, to fuse Carbon requires a temperature of about 1 billion degrees.

A star that is composed of mostly hydrogen is a young star. As a star gets older its hydrogen is consumed and tremendous energy and light is released through fusion. This is a three-step process: (1) two hydrogen nuclei (protons) fuse to form a heavy hydrogen called deuterium and release an electron and 4.04 MeV energy, (2) the deuterium fuses with another hydrogen nucleus (proton) to form a helium-3 and release a neutron and 3.28 MeV energy, and (3) the helium-3 fuses with another helium-3 to form a helium-4 and releases two hydrogens and 10.28 MeV energy.

In stars with central temperatures greater than 600-700 million degrees, carbon fusion is thought to take over the dominant role rather than hydrogen fusion. Carbon fusion can produce magnesium, sodium, neon, or helium. Some of the reactions release energy and alpha particles or protons.

**SKILL 7.4**      **Describe the solar surface as the immediate source of energy for the Earth's surface.**

The vast majority of energy on the Earth's surface is derived from sunlight. However, sunlight is attenuated by the Earth's atmosphere, so that not all this solar energy reaches the planet's surface. Specifically, about 1300 watts are delivered per square meter of Earth, but only about 1000 watts actually reach the surface.

It is easy for us to understand how sunlight warms the land and water on the surface of the Earth. We are similarly familiar with the capture of sunlight by solar cells, which can be used for heating or electricity. However, it is important to understand that sunlight is the basis for many of our other sources of power. This is because sunlight is used to drive photosynthesis, which is the major method by which carbon is fixed by living things. The energy harnessed by plants is used to fuel all heterotrophs further up the food chain. When these life forms (plant or animal) die, they may ultimately be converted to fossil fuels. Thus the petroleum, oil, and other fossil fuels we use as a major power source all originally derived their energy from sunlight.

The Sun and the energy it transfers to the Earth also influence the movement of air and water (i.e., the winds and ocean currents). This is a result of the fact that different areas of the Earth's surface receive varying amounts of energy from sunlight. The movements of warmer and cooler masses of air or water are the source of wind and water currents. Note that this means that wind and hydrologic energy also have their origin in energy from the Sun.

Finally, it is important to recognize that the Sun is a star with its own strong and ever changing magnetic field. These changes in the field are responsible for solar activity such as sunspots, solar flares, and solar wind. These changes in turn affect Earth's climate and sometimes increase or decrease the solar energy that reaches the entire Earth or certain areas. Fluctuations in the solar cycles can also lead to especially low or high temperatures. For instance, during the 17th and 18th centuries there was a period in which very few sunspots were seen. This lasted for about 70 years and coincided with the "Little Ice Age" in Europe, an era of unusually cold temperature throughout the continent. It is not only the extended period of changes in solar activity that effect climate and weather on Earth. For example, it has been observed that stratospheric winds near the equator blow in different directions, depending on changes in the solar cycle.

**SKILL 7.5**      **Understand the sunspot cycle and its possible effects on the Earth's climate.**

A sunspot is a region on the Sun's surface (the photosphere) that is characterized by a lower temperature than its surroundings. The lower temperatures of these areas are caused by intense magnetic activity, which inhibits convection. The surface of the Sun is approximately 5800 K. Sunspots, however, usually demonstrate temperatures around 4000–4500 K. For this reason, sunspots are clearly visible as dark spots on the Sun's surface.

The number of sunspots rises and falls in a periodic manner, with a cycle of approximately 11 years. The 11-year sunspot cycle is related to the 22-year cycle for the reversal of the Sun's magnetic field. During this period, the Sun's magnetic field rotates 360 degrees, causing its poles to switch. The periods of time during which the Sun demonstrates the largest and smallest numbers of sunspots are known as the solar maxima and minima, respectively. The most recent solar maximum was observed in 2001. The Sun began a new cycle with a solar minimum in 2007.

It has recently been determined that areas of the Sun's surface surrounding sunspots increase in brightness as sunspots form. For this reason, sunspot maxima generally correspond to periods of higher solar radiation and solar activity, including increased solar winds, aurorae, magnetic storms, flares, prominences, and non-thermal radio, x-ray and UV emission. Studies estimate that the Sun's radiant energy varies by up to 0.2% between extremes of the sunspot cycle. Because of increased solar irradiation, more energy is delivered to the Earth's atmosphere, and global temperatures may rise. In fact, the combined effects of sunspot-induced changes in solar irradiance and increases in greenhouse gases may best account for the observed rise in global temperatures over the past century.

At solar minima, less energy from the Sun arrives at the Earth's atmosphere. Because the normal sunspot cycle lasts only 11 years, the Earth's climate is not seriously affected at every solar minima. However, solar minima that continue for decades, such as the Sporer, Maunder, and Dalton Minimas, can produce serious cooling effects on Earth. These minima occurred during a period of time known as the "Little Ice Age" (1645-1710). During this time, Europe was struck by famine as record low temperatures were reported. During an earlier Maunder Minimum (1400-1510), a Viking settlement in Greenland was reportedly decimated due to complete failure of crops.

# TEACHER CERTIFICATION STUDY GUIDE

**COMPETENCY 8.0**   Understand the properties, features, and movements of the Earth's moon; the interactions among the Earth, Moon, and Sun (including phases, tides, and eclipses); and the role of technology and exploration in obtaining knowledge about the Earth, Moon, and Sun.

**SKILL 8.1**   Relate surface features (e.g., maria, craters, mountains) of the Earth's moon to events in the history of the moon.

Earth's Moon is covered with mountain ranges, craters, and plains. In the early days of astronomy, the Lunar Plains: vast lava flow plains, were thought to be oceans, and were called maria or mare (Latin derivatives for seas) by Galileo in the 1600's. This name is still used although we know that the plains are actually vast, dry, flat expanses of barren volcanic rock. Because the presence of igneous volcanic rock implies tectonism, we know that at one point in its development the Moon was tectonically active, producing the Lunar Highlands: lunar mountains and the lava plains. However, such activity has long ceased and for all practical purposes, the Moon is tectonically dormant. The Moon has had a tough time since its inception. Meteorite bombardment melted its outermost layer down to 100 kilometer in depth. This initial swarm of meteorite activity ceased long enough for the highlands to form around 4.4 billion years ago. But the calm was not to last. Between 4.2 and 3.9 billion years ago, billions of meteorites pummeled the Moon's surface, blasting huge craters and by 3.8 billion years ago, this continuous assault raised the temperature to the point that the Moon's interior turned molten. This impact-induced melting caused the denser materials to sink downward and the lighter materials to rise to the surface. Volcanic activity continued for 700 million years, lasting until approximately 3.1 billion years ago, when the Moon finally cooled.

**SKILL 8.2**   Describe the relationship between the height of ocean tides and the relative positions of the Earth, Moon, and Sun.

The periodic rise and fall of the liquid bodies on Earth are the direct result of the gravitational influence of the Moon and, to a much lesser extent, the Sun. Tides are produced by the differences between gravitational forces acting on parts of an object.

EARTH SCIENCE

As shown in Newton's Universal Law of Gravitation, the gravitational effect of two bodies is mutually constant and depends largely on the distance and mass between the objects. The side of the Earth that faces the Moon is roughly 4,000 miles (6,400 km) closer to the moon than is the Earth's center. This has the effect of increasing the Moon's gravitational attraction on Earth's oceans and landforms. Although the effect is so small on the mass of the landforms as to be invisible, the effect on the liquid parts is greater. The Moon's gravitational effect causes a bulge to form on both sides of the Earth. If we were able to view such subtle change from outer space, the affected waters would create an elliptical shape, compressing downward at the top and bottom of the planet and extending outward on the sides. This double-bulge effect causes the tides to fall and rise twice a day, and the time of the high and low tides is dependent on the phase of the Moon. Yet not all locations are uniformly affected. The tidal cycle at a particular location is actually a very complicated interaction of the location's latitude, shape of the shore, etc.

Because of its distance from the Earth, the Sun's gravitational effect on tides is only half that of the Moon's. However, when the gravitational effects of both the Sun and Moon join together during a new moon and a full moon phase, the tidal effects can be extreme. During a new moon and a full moon, tidal effects are much more pronounced as the tidal bulges join together to produce very high and very low tides. These pronounced types of tide are collectively known as Spring Tides. During the first and third quarters of the Moon's phases, the Sun's effect is negligible and consequently, the tides are lower. These are called Neap Tides.

**SKILL 8.3** **Describe the relationships between the phases of the Moon and the relative positions of the Earth, Moon, and Sun.**

Just as the Earth follows an orbit around the Sun, the Moon follows an Eastward moving orbit around the Earth. Because the Moon's rotational period matches the Earth's and its period of revolution is 27.3 days (called the sidereal period), this keeps one side of the Moon always facing Earth. The side always facing us is called the Near Side, and the darkened side we never see is called the Far Side.

Phases of the Moon: the apparent change in shape of the Moon caused by the absence or presence of reflected sunlight as the Moon orbits around the Earth. The orbital pattern of the Moon in relation to the Sun and Earth determines the extent of lunar illumination, and consequently, what illuminated shape is presented to the Earth. When the Moon is between the Sun and the Earth, the side facing us is darkened, and we refer to this as a New Moon. The opposite pattern occurs in the second half of the complete Lunar Cycle, when the Moon is fully illuminated and bright in the night sky. This is called a Full Moon. The other phases between these extremes reflect the orbital point of the Moon as it completes its journey around the Earth.

**SKILL 8.4** **Explain how the lunar exploration program has added to our knowledge of the Earth-Moon system.**

Throughout history there have been many manned missions including many "firsts" such as first animal, man on moon, women in space, preventable catastrophe, and first fatal catastrophe. As of 2007 the manned missions have been contained to orbiting around the Earth and landing on the Moon. With new knowledge of propulsion it would be possible to reach Mars with a manned mission. Previously, unmanned missions have used land rovers to collect over 17,000 photo images and collect rock and soil samples.

| | |
|---|---|
| **Soyuz TM-32** | The Russian Soyuz TM-32 was to be kept at the International Space Station as a lifeboat, and the crew that brought it returned to Earth on the Soyuz TM-31 stored there. |
| **Vostok 6** | The Russian Vostok 6 was to continue experiments for joint spaceflights and also observe the effect of space travel on the female body. |
| **Apollo-Soyuz** | This mission involved a docking of ships between the American Apollo and Russian Soyuz to develop techniques for emergency rescues, as well as to perform some experiments. |
| **Challenger** | The American Challenger explosion occurred on the tenth mission of this space shuttle. |
| **Vostok 1** | This Russian mission was the first manned spaceflight in history. |
| **Apollo 13** | This American mission was to gather information and pictures from the Moon. There were several problems that resulted in a dangerous explosion. |
| **Apollo 11** | This American mission was the first lunar landing that also brought the first man on the surface of the Moon. |
| **Shenzhou 5** | This was the People's Republic of China's first manned flight. Liwei Yang was the first Chinese man in space. |
| **Voskhod 1** | In this mission, the USSR launched the first space flight with more than one person aboard. This was also the first flight without spacesuits. |

**COMPETENCY 9.0** Understand the scale and organization of the solar system, the role of gravity in the solar system, characteristics of the bodies within the solar system, and physical and mathematical models that describe these objects and their real and apparent motions.

**SKILL 9.1** Describe the characteristics (e.g., size, density, surface temperature) of the planets.

**The Inner Planets**

| MERCURY | |
|---|---|
| Average Distance from the Sun | 0.387 AU ($5.79 \times 10^7$ km) |
| Density | 5.44 g/cm$^3$ |
| Diameter (equatorial) | 4878 km (0.38 of Earth) |
| Mass | $3.31 \times 10^{23}$ kg (0.055 of Earth) |
| Atmosphere | Only trace gasses only of H, He, Na, K |
| Axial Rotation Period | 58.6 days |
| Solar Revolution Period | 87.9 days |
| Axial Tilt | 0° |
| Surface Temp | -173°C to 330°C |
| Moons | None |
| Composition | Silicate & Iron rocks |

| VENUS | |
|---|---|
| Average Distance from the Sun | 0.723 AU ($1.082 \times 10^8$ km) |
| Density | 5.3 g/cm$^3$ |
| Diameter (equatorial) | 12,104 km (0.95 of Earth) |
| Mass | $4.87 \times 10^{24}$ kg (0.82 of Earth) |
| Atmosphere | Runaway Greenhouse effect 96.5% $CO_2$, 3.5% N. |
| Axial Rotation Period | 243.01 days |
| Solar Revolution Period | 224.68 days |
| Axial Tilt | 177° (Makes the planet appear as if it was rotating clockwise) |
| Surface Temp | 472°C |
| Moons | None |
| Composition | Unknown, crustal materials believed to be similar to Earth |

| EARTH | |
|---|---|
| Average Distance from the Sun | 1.00 AU (1.495 x $10^8$ km) |
| Density | 5.497 g/cm$^3$ |
| Diameter (equatorial) | 12,756 km |
| Mass | 5.976 x $10^{24}$ kg |
| Atmosphere | 78% Nitrogen, 21% Oxygen |
| Axial Rotation Period | 24.00 hours |
| Solar Revolution Period | 365.26 days |
| Axial Tilt | 23.5° |
| Surface Temp | -50°C to 50°C |
| Moons | Just one, the Moon |

| MARS | |
|---|---|
| Average Distance from the Sun | 1.523 AU (2.279 x $10^8$ km) |
| Density | 3.94 g/cm$^3$ |
| Diameter (equatorial) | 6796 km (0.53 of Earth) |
| Mass | 0.6424 x $10^{24}$ kg (0.1075 of Earth) |
| Atmosphere | 95% $CO_2$ |
| Axial Rotation Period | 24.61 hours |
| Solar Revolution Period | 686.9 days |
| Axial Tilt | 23° 59' |
| Surface Temp | -140°C to 20°C |
| Moons | Deimos and Phobos. They may actually be captured asteroids because of their irregular shape. |
| Composition | Identical to Earth's |

**The Outer Planets**

| JUPITER | |
|---|---|
| Average Distance from the Sun | 5.202 AU (7.783 x $10^8$ km) |
| Density | 1.34 g/cm$^3$ |
| Diameter (equatorial) | 142,900 km (11.20 of Earth) |
| Mass | 1.899 x $10^{27}$ kg (317.83 of Earth) |
| Atmosphere | Hydrogen, Helium, and Ammonia |
| Axial Rotation Period | 9.83 hours |
| Solar Revolution Period | 11.867 years |
| Axial Tilt | 3.5° |
| Surface Temp | 29,727 °C at surface<br>-120°C in atmosphere |
| Moons | 47, of which only 4 are significant. These are the **Galilean Moons:** Io, Europa, Ganymede, and Callisto. |
| Composition | No surface, gaseous atmosphere |

| SATURN | |
|---|---|
| Average Distance from the Sun | 9.538 AU (14.27 x $10^8$ km) |
| Density | 0.69 g/cm$^3$ |
| Diameter (equatorial) | 120,660 km (9.42 of Earth) |
| Mass | 5.69 x $10^{26}$ kg (95.17 of Earth) |
| Atmosphere | Hydrogen and Helium |
| Axial Rotation Period | 10.65 hours |
| Solar Revolution Period | 29.461 years |
| Axial Tilt | 26° 24' |
| Surface Temp | Unknown at surface and -180° in atmosphere |
| Moons | 17 total, of which Titan is the largest and most significant. |
| Composition | No surface, gaseous atmosphere |

| URANUS | |
|---|---|
| Average Distance from the Sun | 9.538 AU (14.27 x $10^8$ km) |
| Density | 1.29 g/cm$^3$ |
| Diameter (equatorial) | 51,118 km (4.01 of Earth) |
| Mass | 8.69 x $10^{25}$ kg (14.54 of Earth) |
| Atmosphere | Hydrogen, Helium, and Methane |
| Axial Rotation Period | 17.23 hours |
| Solar Revolution Period | 84.013 years |
| Axial Tilt | 97° 55' |
| Surface Temp | Unknown at surface -220° in atmosphere |
| Moons | 15 total, of which Oberon is the largest at 1,500 km diameter. |
| Composition | Unknown, frozen gasses |

| NEPTUNE | |
|---|---|
| Average Distance from the Sun | 30.061 AU (44.971 x $10^8$ km) |
| Density | 1.66 g/cm$^3$ |
| Diameter (equatorial) | 49,500 km (3.88 of Earth) |
| Mass | 10.30 x $10^{26}$ kg (17.23 of Earth) |
| Atmosphere | Hydrogen, Helium, and Methane |
| Axial Rotation Period | 16.05 hours |
| Solar Revolution Period | 164.793 years |
| Axial Tilt | 28° 48' |
| Surface Temp | Unknown at surface -216°C in atmosphere |
| Moons | 8 total, 2 of significance, Triton and Nereid. |
| Composition | Unknown, mostly ice |

**EARTH SCIENCE**

| PLUTO | |
|---|---|
| Average Distance from the Sun | 39.44 AU (59.00 x $10^8$ km) |
| Density | 2.0 g/cm$^3$ |
| Diameter (equatorial) | 2,300 km (0.19 of Earth) |
| Mass | 1.2 x $10^{22}$ kg (0.002 of Earth) |
| Atmosphere | Nitrogen and Methane |
| Axial Rotation Period | 9.3 days |
| Solar Revolution Period | 247.7 years |
| Axial Tilt | 122° |
| Surface Temp | -230°C |
| Moons | One, Charon |
| Composition | Frozen nitrogen and rock |

**SKILL 9.2**     **Identify the relative sizes, distances, tilts, and positions of the planets.**

The solar system is divided into two sections: the inner and outer planets. The composition of the inner planets reflects the attraction of the heavier elements by the Sun. The outer planets' composition reflects the lighter, less dense elements not attracted as much by the Sun's gravitational mass. Heavy elements sink inward to form the core. Lighter elements form the atmosphere. Our solar system consists of the Sun, planets, comets, meteors, and asteroids.

| Comparison of the Basic Characteristics of the Inner and Outer Planets: | |
|---|---|
| **INNER PLANETS** | **OUTER PLANETS** |
| Referred to as the Terrestrial Planets | Called the Gas Giants or Jovian Planets |
| Similar to density of Earth. | Except for Pluto, they are very large in size. |
| Also referred to as the "Rocky Planets." | Primarily composed of gas. |
| Relatively small in size. | Less dense than Earth. |
| Spin slowly on their axis. | Rotate rapidly on their axis. |
| Few, if any, moons. | Lots of moons. |
| Mercury | Jupiter |
| Venus | Saturn (Ringed) |
| Earth | Uranus |
| Mars | Neptune |
| + Asteroid Belt | Pluto (Rocky) |

**SKILL 9.3**         **Identify the position of the planets on the ecliptic.**

Order of the Planets is: Mercury, Venus, Earth, Mars, Asteroid Belt, Jupiter, Saturn, Uranus, Neptune, and Pluto. Memory aid for remembering the order of the planets: My Very Educated Mother Just Served Us Nine Pies.

All the planets revolve around the Sun, and all of the planets, with the exception of Venus, rotate on their axis in the same direction. Venus has a retrograde motion; it rotates backwards. Except for Pluto, all the planets follow roughly the same elliptical orbital planes around the sun. Neptune and Pluto occasionally change places in the order. Pluto's orbit is very erratic compared to the other planets and sometimes it carriers Pluto inside of Neptune's orbit. The asteroid belt is located between Mars and Jupiter and may be the remnants a planet crushed by the massive gravitational force of Jupiter. As of the Summer of 2006, Pluto's status as a planet was being reconsidered. Because of its abnormalities some scientists are suggesting that it isn't truly a planet, but instead just a celestial body.

**SKILL 9.4**         **Describe the origin and properties of comets and meteors.**

Comets: small icy bodies that orbit the Sun and produce a glowing tail of gas and dust as the comet nears the Sun. Comets are not planets.

Asteroids: small, rocky worlds that orbit the Sun. Asteroids are sometimes called the minor planets.

Meteoroids: a meteor in space. Meteoroids are often asteroid fragments, but they do not have a specific orbit. Meteroids can become either meteors or meteorites depending on whether they burn up or make it through the atmosphere.

Meteors: small fragments of rock debris that turn incandescent and burn up upon entering the Earth's atmosphere.

Meteorites: meteor fragments that do not burn up in the atmosphere and actually strike a planet.

**SKILL 9.5**     **Use the apparent motion of celestial objects to infer solar system models (i.e., geocentric, heliocentric).**

**Geocentric Model**: the original model of universal object arrangement in which the universe was Earth centered and all other stellar objects rotated around the Earth. The Sun followed a track around the outer edge of the rotational sphere. The ancient Greeks later modified this model into the Greek Celestial Model (Aristotelian Model).

The ancient **Greek Celestial Model (Aristotelian Model)** was also geocentric. The Earth is still at the center of the universe but the planets, Sun, and stars arer all on different spheres orbiting around the Earth. God (heaven) is on the outermost sphere. This model was widely accepted because it was based on the philosophical teachings of Aristotle.

**Ptolemaic Model:** The Alexandrian Greek Astronomer Ptolemy, observing the planetary motions in 140 A.D., realized that reality didn't fit the Greek model well. He modified the Greek Celestial model to fit his observations while still retaining a geocentric orientation. Ptolemy's model had the moon orbiting around the Sun in an Epicyclic motion. An **epicycle is a retrograde**: backward motion, in which and object spins or moves clockwise.

**Heliocentric Model**

Nicolaus Copernicus (1507 A.D.)- Based solely on his extensive planetary observations, Polish monk Nicolaus Copernicus, in his paper *"Revolutions of the Heavenly Spheres,"* triggered a complete revision of human thought by publicly proposing that the universe is heliocentric, not geocentric. He also claimed that the Earth's axis was tilted in relation to the Sun, using retrograde motion, only in a different way.

Copernicus knew that his theory would be highly controversial. Out of fear of excommunication by church authorities, he didn't publish his paper until near the end of his life. As he expected, his paper raised a storm of argument between scholars and the church authorities. The debate between Copernicus' supporters and detractors grew so heated as to lead to bloodshed on several occasions.

Tycho Brahe (1572 A.D.)- Spurred on by the growing interest in science, Danish nobleman Tycho Brahe rejected a career as a lawyer and politician, and devoted his life to astronomy and mathematics. In 1572, Tycho, as he is usually called, observed a brilliant new star in the sky and, like other classically trained astronomers of the day, he was puzzled by its appearance. This new star did not fit into the Aristotelian scheme of the heavens, which held that the starry regions were perfectly complete. This meant that the star must be between the moon and the Earth. However, by measuring the parallax of motion, his calculations showed that the star must be further away, challenging the Ptolemaic concepts.

His conclusions were published in a small book *"De Stella Nova* (the New Star)" in 1573 and attracted the attention of the King of Denmark, who offered him funds to build a planetary observatory on the Danish isle of Hveen. Tyco's greatest contribution to astronomy was not theoretical. Instead, his true value to science lay in his observations.

For over 20 years Tycho constructed new and also greatly improved existing instruments. He studied the motion of over 700 stellar objects, collecting the most complete data sets of astronomical observations since the early Egyptian times. Amazingly, he did this with the naked eye, as the telescope wasn't invented until almost a hundred years after his death. Tycho hired several mathematicians and astronomers to assist him in collecting and compiling the data. Among these individuals was Johannes Kepler, who would continue Brahe's work following Tycho's death in 1610.

Johannes Kepler (1600 A.D.)- In collating Tycho Brahe's data sets, Johannes Kepler discovered that the data wouldn't fully support the Copernican model. In an effort to try and make the data fit, he postulated what are now known as Kepler's Laws of Planetary Motion.

The significance of Kepler's Laws is that they overthrew the ancient concept of uniform circular motion, which was a major support for the geocentric argument. Although Kepler postulated three laws of planetary motion, he was never able to explain why the planets move along their elliptical orbits, only that they did.

Galileo Galilei (1633 A.D.)- One of the truly brilliant minds of his time, Galileo Galilei was an Italian astronomer and inventor who built his first telescope in 1609. Contrary to popular myth, Galileo didn't invent the telescope but used existing plans and gradually improved upon the design.

The major significance of Galileo's observations is that they totally disproved the geocentric model of the universe. In doing so, Galileo totally destroyed the Aristotelian vision of the universe that had been the basis of astronomical and philosophical though for over 2,000 years.

Galileo's destruction of the Aristotelian theory of the universe created a large problem. If Aristotle was correct, what was the truth? People began to openly question blind belief in religious explanations and sought the answer to their questions through science.

**SKILL 9.6      Apply Kepler's laws to describe and predict the motions of the planets.**

Kepler's First Law of Planetary Motion: A planet can't travel in a circle. A planet travels an elliptical path, with the Sun at one foci point. Ellipse: a geometric figure drawn around two points, called foci.

Kepler's Second Law of Planetary Motion: Planets must sweep out equal areas, at equal amounts of time, on these elliptical paths. When a planet is nearer to the Sun on the planet's elliptical path, it must move faster to sweep over the same area.

Kepler's Third Law of Planetary Motion: A planet's orbital period squared is proportional to its average distance from the sun cubed ($a^3 = p^2$). *Example:* Jupiter's average distance to the Sun is 5.20 AU. If a equals 5.20 AU, then $a^3$ equals 140.6 AU. Therefore, the orbital period of Jupiter must be the square root of 140.6, which equals a period of about 11.8 light-years.

The significance of Kepler's Law is that it overthrew the ancient concept of uniform circular motion, which was a major support for the geocentric arguments. Kepler suggested that the eccentricity of the planetary motion could be mathematically summarized as $e=c/a$.

The two major points on the elliptical orbit are the Perihelion and Aphelion.

- Perihelion: the closest point of approach to the Sun on the elliptical path. This is the point at which the planet travels the fastest in keeping with Kepler's 2$^{nd}$ Law.
- Aphelion: the farthest point away from the Sun on the elliptical path. This is the point at which the planet travels the slowest in keeping with Kepler's 2$^{nd}$ Law.

## TEACHER CERTIFICATION STUDY GUIDE

**COMPETENCY 10.0**  Understand stars, their motions and life cycles, and the methods and technology used to study them.

**SKILL 10.1**  Compare types of telescopes (e.g., optical, radio, infrared, ultraviolet) and the ways in which they are used to acquire information on star characteristics.

**Telescopes and Light**

There are two main styles of optical telescopes: refractor and reflector.

**Refractor Telescopes**

**Refraction**: the bending of light. Example: Put a straw in a clear glass of water. Now look at the straw through the side of the glass. It will appear to bend at the point where the straw enters the water.

**Refractor Telescope**: an optical device that makes use of lenses to magnify and display received images.  Professional astronomers do not use Refractor telescopes because of two main problems: first, the telescopes are affected by chromatic aberrations which make it difficult to focus on the stars, and second, because they rely solely on lenses, the telescopes have inherent weight and size restrictions.  Chromatic Aberrations are a problem because as the lenses split the light into its chromatic components; each color has a different focal point based on its wavelength. This causes details on the image to blur.  To compensate for this problem lenses with different refractive indexes (the degree of bending) are sandwiched together in order to compensate for the aberrations and properly focus on the image. However, this compensation method increases the size and weight of the telescope.

**Reflector Telescopes**

**Reflection**: the re-emission of light off of an object struck by the light.
Example: Look at yourself in the mirror.  What your eyes see is the re-emission of light waves that have struck you and the mirror.

**Reflector Telescope**: an optical device that makes use of a mirror or mirrors, to reflect light waves to an eyepiece (an ocular), thereby eliminating chromatic aberrations.  There are different types of reflector telescopes; some use mirrors only, and others make use of both lenses and mirrors.  The <u>objective</u> is the primary focus mirror in a reflector telescope and is usually curved. Any other mirrors in the telescope are referred to as secondary, tertiary, etc. depending on the number of mirrors present.  On both refractor and reflector telescopes, the eyepiece is called the ocular.

**EARTH SCIENCE**

The most common style of reflector telescope used is the Schmidt-Cassegrain. The two major advantages of a reflector telescope over a refractor style telescope are:

- No chromatic aberrations, which results in a much clearer and detailed image than in a refractor telescope.

- Their smaller size makes them easier to use, and they weigh less because they don't rely on the heavy, thick glass lenses required in a refractor type.

However, there are still limits to the size of a reflecting telescope. They are very expensive and it is an extremely difficult and slow process to grind the mirror to the proper specifications. The larger the mirror or mirrors required to view distant objects, the more likely it is that there will be **imperfections** in the mirror. Weight also becomes an issue depending on the size of the mirror required. The heavier and thicker the mirror required, the greater the chance that the mirror will sag, slightly distorting the image. Also, the mirror will heat and cool unevenly, causing further distortion.

### Active Optics

**Active Optics:** a type of optical device that is composed of hexagonal pieces of mirror whose positions are controlled by a computer. Also referred to as **Active Telescopes**.

Collectively, smaller pieces of mirror weigh less than a single large mirror and, more importantly, they generally do not suffer from sagging problems. Small hexagonal shaped pieces of mirror are arranged next to each other to form a larger reflection surface. Computer-controlled thrusters mounted underneath the pieces control the mirror position and focus. Employing the smaller pieces which work in conjunction eliminates sagging and the uneven heating and cooling problems found in extremely large, single mirror type telescopes.

Example: The Keck Telescope in Hawaii is able to have a 10-meter diameter reflective surface through the use of active optics. Similarly, the Hobby-Eberly Telescope is composed of 91 hexagons has an 11-meter diameter reflective surface, making it the largest telescope on Earth.

## New Generation Telescopes

The **Hubble Space Telescope** is named in honor of the American astronomer Edwin Hubble who proved the theory of an expanding universe. The Hubble Space Telescope, although only possessing a 2.4-meter diameter reflective surface, isn't affected by atmospheric constraints, and as a result, it provides a much clearer, higher resolution image of stellar objects than is possible through the use of an Earth-based telescope. Two companies competed in the design phase of manufacturing the telescope, and working models of each design were constructed. However, shortly after being positioned in space, the mirror on the winning design was discovered to have imperfections that produced a great deal of distortion in the received images. Initially, the fix for this problem seemed simple; put the losing contractor's design up in space in place of the flawed telescope. But a problem quickly arose. When NASA scientists contacted the second company, they were told that the losing model no longer was available. In any event, NASA eventually figured out a means to correct the imperfections and space shuttle astronauts successfully affected the repairs.

## CCD-Charged Coupled Devices

A **CCD** is a camera plate made up of thousands of tiny pixels. The pixels carry a slight electrical charge and when photons strike a pixel, electrons are released. The release of electrons causes a flow of current through an attached wire, and this current is detected by a computer chip and used to construct images based on the number of strikes. The number of strikes also shows the intensity of the received image.

CCD cameras have a wide range of applications besides astronomy. This type of technology is being successfully employed in many of the newest, high-resolution cameras available to the general public.

## Radio Telescopes

Variances in radio waves received from space can be translated into usable astronomical data. The advantages offered by use of radio telescopes are many: it's cheaper to build a radio telescope than optical telescopes, they can operate 24 hours a day and be built just about anywhere on Earth, and they open up an entirely new window of space investigation. But they initially had one major disadvantage: the useable radio waves received from space were not overly abundant and generally very weak, and you needed a huge receiving dish to detect the signals. To overcome these problems, scientists developed a technique called Radio Interferometry.

**Radio Interferometry**: a method of amplifying weak radio waves by lying out, in a Y-shaped pattern, a series of small radio telescopes all pointed at the same point in the sky. The telescopes add their received signals together to form an overlay of signals. Computers control the angle of incidence and correlate the incoming signals. This improves resolution and limits the size of the radio telescope dish needed for a single unit.

## Black Body Radiation

When objects are heated, they give off properties of light, and you observe different colors at different temperatures. Example: Heat a needle. The color of the needle will go from red to white. The hotter you make it, the more the color will vary.

We measure the intensity of the light by the ratio of the apparent presence or absence of colors. **Intensity**: the amount of light contained in a space. Intensity varies by distance. The further away you are, you will see a drop in intensity equal to 1 over the distance squared ($1/d^2$). Intensity gives off a continuous amount of color, but the intensity of the colors seen varies in accordance with the temperature of the object.

Stars follow the same principles of emission. We observe the intensity of the stars by using a red and blue filter on a photon counter mounted to a telescope. The red and blue ratio determines the color index. From the color index, we can determine the temperature, size, properties, and material composition of the star. However, appearances can be deceiving. The hotter the object the bluer it will appear. This is because there is an inverse relationship between temperature and wavelength. Wavelength decreases as temperature increases.

**SKILL 10.2    Identify methods and uses of spectroscopy.**

**Spectroscopy**: the study of the spectrum.

**Spectra**: the emitted electromagnetic or visible light line pattern produced by an object. There are three different types of spectra: Continuous, Absorption, and Emission.

**Continuous Spectrum**: light from a glowing source. Example: A prismatic rainbow. All the colors that comprise the white light are received.

**Absorption Spectrum**: a continuous colored spectrum, interrupted by dark lines.

**Emission Spectrum**: a continuous, dark spectrum interrupted by lines of color. The lines of color appear to be slices of the continuous spectrum.

The received spectra are unique to each element and act as the "fingerprints" of the element. By observing the spectra, we can determine a star's temperature and elemental composition. If the lines are shifted, we can determine how far and at what speed the star is moving. We do this by comparing the star's spectral emissions to known data for various Earthly elements.

**SKILL 10.3  Identify types of stars (e.g., pulsars, Cepheid variables) and their characteristics.**

A star's life cycle depends on its initial mass. Red stars have a small mass. Yellow stars have a medium mass. Blue stars have a large mass. Large mass stars consume their hydrogen at a faster rate and have a short life cycle in comparison to small mass stars that consume their hydrogen at a much slower rate. All stars eventually convert a large percentage of their hydrogen to heavier atoms and begin to die. However, just as their mass determines the length of their life, it also determines the pattern they follow in the last stages of their existence.

**Lower Main Sequence Stars**

When small and medium mass stars (such as the Sun) consume all of their hydrogen, their inner cores begin to cool. The stars begin to consume the heavier elements produced during fusion (carbon and oxygen) and the star's shell tremendously expands outward, causing the star to become a **Giant Star**: a large, cool, extremely luminous star 10 to 100 times the diameter of the Sun. Example: In roughly 4.6 billion years from now our Sun will become a giant star. As it expands, its outer layers will reach halfway to Venus.

The dying Giant gives off thermal pulses approximately every 200,000 years, throwing off concentric shells of light gasses enriched with heavy elements. As it enters its last phases of the life cycle, its depleted inner core begins to contract, and the Giant becomes a **White Dwarf Star**: a small, slowly cooling, extremely dense star, no larger than 10,000 km in diameter.

The final phase of a lower main sequence star life cycle can take two paths. After a few billion years most main sequence white dwarfs completely burn out to become **Black Dwarfs**: cold, dead stars. However, if a White Dwarf is part of a **Binary Star**: two suns in the same solar system, instead of slowly cooling to become a Black Dwarf, it may capture hydrogen from its companion star. If this happens, the temperature of the White Dwarf soars. When it reaches approximately 10 million °C, a nuclear explosion occurs, creating a **Nova**: a sudden brightening of a lower main sequence star to approximately 10,000 times its normal luminosity; caused by the explosion of the star. A nova reaches its maximum brightness in a short time (one or two days) and then gradually dims as the gasses and cosmic dust cool.

## Upper Main Sequence Stars

The initial sequence of the high mass, upper main sequence stars is identical to the lower mass stars (Planetary Nebulae to Protostar). However, if the protostar accretes enough material, it forms a **Blue Star**. When a Blue Star has consumed all of its hydrogen it expands outward, but on a much larger scale then experienced by a lower mass star. It becomes a **Supergiant Star**: an exceptionally bright star, 10 to 1000 times the diameter of the Sun.

The Supergiant's now depleted core cannot support such a vast weight and collapses inward, causing its temperature to soar. When it reaches roughly 599 million °C it implodes and then explodes, creating a **Supernova**: the massive explosion of an upper main sequence Supergiant star caused by the detonation of carbon within the star.

A supernova releases 90% of its shattered mass into space, becoming planetary nebulae from which the life cycle may begin anew. The other 10 percent, the core of the star, is blown inward, becoming a **Neutron Star**: a very small, 10 km diameter core of a collapsed Supergiant star that rotates at a high speed (60,000 rpm) and has a strong magnetic field (1012 Gauss).

A neutron star may capture gas from space, a companion star, or a nearby star and become a **Pulsar**: a neutron star that emits a sweeping beam of ionized gas radiation. As the pulsar rotates, the beams of light sweep into space similar to a beacon from a lighthouse. Since first discovered in 1967, over 350 pulsars have been catalogued.

The alternate product of a supernova is a **Black Hole**: a volume of space from which all forms of radiation cannot escape. Black Holes are created when a Supergiant star with a mass roughly 3 times that of the Sun implodes. The inner core of the star is compacted by the supernova into a **Singularity**: an object of zero radius and infinite density. A singularity is difficult to picture. A zero radius implies an object with size less than an electron, but also possessing a density that precludes the escape of all radiation including light. Although a singularity has yet to be detected, theoretically, they exist in and cause the effects exhibited by Black Holes.

A **Cepheid** variable is known for its regularity. Cepheid stars are most often giant yellow stars that pulse regularly by expanding and contracting, resulting in changes in its luminosity. Cepheid stars range from 103 to 104 times more luminous than the Sun. There is regularity when observing its period of variability and its absolute luminosity. Because of this, a Cepheid variable can be used as a standard to determine the distance to its galaxy. Distances found with this method are among the most accurate.

**SKILL 10.4    Use the H-R diagram to analyze the life cycles of stars.**

**The Characteristics of Stars**

Stars are not all alike. Their energy outputs vary from 111,000th of the energy to 100,000 times the energy of Earth's Sun. The laws of physics tell us that the more energy an object has, the hotter it is. They also relate color to temperature. Therefore, by observing the color of a star, we get information about its temperature.

**The Hertzsprung-Russell Diagram**

In 1913, American astronomer Henry Norris Russell and Danish astronomer Ejnar Hertzsprung theorized that the energy emitted by a star is directly related to the star's color. Their supporting graph is now known as the **Hertzsprung-Russell Diagram (H-R Diagram):** a graph that shows the relationship between a star's color, temperature, and mass.

On a H-R Diagram the majority (90%) of the stars plotted form a diagonal line called the **Main Sequence**: the region of the H-R Diagram running from top left to bottom right. Hot, blue stars are at the top left while cooler, red stars are at the bottom right. The middle of the main sequence contains yellow stars, like the Sun.

Star mass is also shown on the H-R diagram, increasing from the bottom to the top of the main sequence. The stars at the bottom right have masses about one-tenth of that of Earth's Sun, and the masses increase until you reach the top left where there are stars with masses ten times greater than that of the Sun.

The truly large stars are called **Supergiant Stars**: exceptionally massive and luminous stars 10 to 1,000 times brighter than the Sun, and the smallest stars are called **Dwarf Stars**: dying stars that have collapsed in size. Although small in size, dwarf stars are extremely dense. In contrast, Supergiants are extremely large, but they may be less dense than the Earth's outer atmosphere.

We can also use the H-R Diagram to illustrate the life cycle of stars.

**SKILL 10.5**     **Analyze stellar life cycles to understand the formation and initial development of the solar system.**

**The Life Cycle of Stars**

Stars form in **Planetary Nebulae**: cold clouds of dust and gas within a galaxy, and they go through different stages of development in a specific sequence. This theory of star development is called the **Condensation Theory**.

**Sequence of Development**

In the initial stage, the diffuse area of the nebula begins to shrink under the influence of its own weak gravity. The cloud-like spheres condense into a knot of gasses called a **Protostar**. The original diameter of the protostar is many times greater than the diameter of our solar system, but gravitational forces cause it to continue to contract. This compression raises the internal temperature of the protostar.

When the protostar reaches a temperature of around 10 million °C (18 million °F), nuclear fusion begins, which stops the contraction of the protostar and changes its status to a star.

**Nuclear Fusion**: the process in which hydrogen atoms fuse together to form helium atoms, releasing massive amounts of energy during the fusion reaction. It's the fusion of atoms, not combustion, which causes the star to shine.

A star's life cycle depends on its initial mass. Red stars have a small mass. Yellow stars have a medium mass. Blue stars have a large mass. Large mass stars consume their hydrogen at a faster rate and have a short life cycle in comparison to small mass stars that consume their hydrogen at a much slower rate. All stars eventually convert a large percentage of their hydrogen to heavier atoms and begin to die. However, just as their mass determines the length of their life, it also determines the pattern they follow in the last stages of their existence.

**Lower Main Sequence Stars**

When small and medium mass stars (such as the Sun) consume all of their hydrogen, their inner cores begin to cool. The stars begin to consume the heavier elements produced by fusion (carbon and oxygen) and the star's shell expands outward, causing the star to become a **Giant Star**: a large, cool, and extremely luminous star 10 to 100 times the diameter of the Sun. Example: In roughly 4.6 billion years from now our Sun will become a giant star. As it expands, its outer layers will reach halfway to Venus.

The dying Giant gives off thermal pulses approximately every 200,000 years, throwing off concentric shells of light gasses enriched with heavy elements. As it enters its last phases of the life cycle its depleted inner core begins to contract, and the Giant becomes a **White Dwarf Star**: a small, slowly cooling, extremely dense star, no larger than 10,000 km in diameter.

The final phase of a lower main sequence star's life cycle can take two paths. After a few billion years, most main sequence white dwarfs completely burn out to become **Black Dwarfs**: cold, dead stars. However, if a White Dwarf is part of a **Binary Star:** two suns in the same solar system, instead of slowly cooling to become a Black Dwarf, it may capture hydrogen from its companion star. If this happens, the temperature of the White Dwarf soars. When it reaches approximately 10 million °C, a nuclear explosion occurs, creating a **Nova:** a sudden brightening of a lower main sequence star to approximately 10,000 times its normal luminosity; caused by the explosion of the star. A nova reaches its maximum brightness in a short time (one or two days) and then gradually dims as the gasses and cosmic dust cool.

**Upper Main Sequence Stars**

The initial sequence of the high mass, upper main sequence stars is identical to the lower mass stars (Planetary Nebulae to Protostar). However, if the protostar accretes enough material, it forms as a **Blue Star**. When a Blue Star has consumed all of its hydrogen it, too, expands outward, but on a much larger scale then experienced by a lower mass star. It becomes a **Supergiant Star**: an exceptionally bright star, 10 to 1000 times the diameter of the Sun.

The Supergiant's now depleted core cannot support such a vast weight and collapses inward, causing its temperature to soar. When it reaches roughly 599 million °C, it implodes and then explodes, creating a **Supernova**: the massive explosion of an upper main sequence Supergiant star caused by the detonation of carbon within the star.

A supernova releases more energy than Earth's Sun will produce in its entire life cycle. The luminosity of a supernova is as bright as 500 million Suns. For example: In 1054, Chinese astronomers recorded the sudden appearance of a new star in what is now known as the Taurus Constellation. Bright enough to be seen during the daytime for over a month, it remained visible for two years.

The explosive release of energy in a supernova is so great (1,028 megatons of TNT) as to literally blow the atomic nuclei of the carbon to bits. The shattered mass is accelerated outward at nearly the speed of light (300,000 km/sec. or 186,000 mph). Ninety percent of the shattered mass scatters into space, becoming planetary nebulae from which the life cycle may begin anew. The other 10 percent, the core of the star, is blown inward, becoming a **Neutron Star**: a very small (10 km diameter) core of a collapsed Supergiant star that rotates at a high speed (60,000 rpm) and has a strong magnetic field (1012 Gauss).

A neutron star may capture gas from space, a companion star, or a nearby star and become a **Pulsar**: a neutron star that emits a sweeping beam of ionized-gas radiation. As the pulsar rotates, the beams of light sweep into space similar to a beacon from a lighthouse. Since first discovered in 1967, over 350 pulsars have been catalogued.

The alternate product of a supernova is a **Black Hole**: a volume of space from which all forms of radiation cannot escape. Black Holes are created when a Supergiant star with a mass roughly 3 times that of the Sun implodes. The inner core of the star is compacted by the supernova into a Singularity: an object of zero radius and infinite density. A singularity is difficult to picture. Zero radiuses imply objects with size less than an electron, but also possessing a density that precludes the escape of all radiation including light. Although a singularity has yet to be detected, theoretically, they exist in and cause the effects exhibited by Black Holes.

**Formation of Earth and the Solar System**

Most cosmologists believe that the Earth is the indirect result of a supernova. The thin cloud (planetary nebula) of gas and dust from which the Sun and its planets are formed, was struck by the shock wave and remnant matter from an exploded star(s) outside of our galaxy. In fact, the stars manufactured every chemical element heavier than hydrogen.

The turbulence caused by the shock wave caused our solar system to begin forming as it absorbed some of the heavy atoms flung outward from the supernova. In fact, our solar system is composed mostly of matter assembled from a star or stars that disappeared billions of years ago.

Around five billion years ago our planetary nebulae spun faster as it condensed and material near the center contracted inward forming a proto-sun. As more materials came together, mass and gravitational attraction increased, pulling in more mass. This cycle continued until the mass reached the point that nuclear fusion occurred and the Sun was born.

Concurrently, the proto-sun's gravitational mass pulled heavier, denser elements inward from the clouds of cosmic material that surrounded it. These elements eventually coalesced through the process of **accretion**: the clumping together of small particles into large masses (the planets of our solar system).

The period of accretion lasted approximately 50 to 70 million years, ceasing when the proto-sun experienced nuclear fusion to become the Sun. The violence associated with this nuclear reaction swept through the inner planets, clearing the system of particles, and ending the period of rapid accretion.

The closest planets (Mercury, Venus, and Mars) received too much heat and consequently did not develop the planetary characteristics to support life. The farthest planets did not receive enough heat to sufficiently coalesce the gasses into solid form. Earth was the only planet in the perfect position to develop the conditions necessary to maintain life.

**COMPETENCY 11.0**     **Understand evidence regarding the size, structure, scale, and motions of the universe, the Milky Way galaxy, and the solar system.**

**SKILL 11.1**     **Describe the evidence regarding the location of the solar system within the Milky Way galaxy.**

**The Milky Way Galaxy**

**Galaxy**: a large group (billions) of stars held together by the attraction of mutual gravitation.

If you look into the night sky you may see a ribbon of stars packed so densely together that it appears to be a star lit cloud. You are looking at **the Milky Way Galaxy**, the galaxy in which our solar system is located. This ribbon of brilliance in the night is actually a collection of over 180 billion stars and a huge volume of interstellar dust and gasses.

You may also notice that there are groups of stars that appear closer together. These are **Globular Clusters**: a tightly grouped, high concentration of stars.

These spherically arranged masses of stars are believed to be the oldest stars in the galaxy, approximately 10-20 billion years old. Each of the clusters contains between 10,000 to 1,000,000 individual stars and virtually no interstellar dust.

Although not the prevalent form, a spherical arrangement of stars is not uncommon. The Milky Way Galaxy is a **Spiral Galaxy**: a grouping of stars arranged in a thin disk. It is arranged in a spiraling geometric pattern, which has a central pivot point (nucleus) and arms radiating outward on which stars rotate around the nucleus, somewhat suggestive of the shape of a pinwheel. It is approximately 100,000 light-years in diameter and 2,000 light-years thick at the center, decreasing to 1,000 light-years at the edges.

Our Sun is located on one of the arms of the galaxy, roughly 30,000 light-years from the center of the galaxy, midway between the upper and lower edges.

The Sun rotates around the center of the galaxy at a speed of 250 km/s, and makes one full rotation every 200 million years. This rotational pattern means that in the estimated 4.6 billion years of Earth's existence, our Sun has completed 23 rotations.

**SKILL 11.2** **Identify historical methods of inferring the size, structure, and motions of the galaxy and the solar system (e.g., star observations and counts).**

**Constellation:** a region of the night sky in which a group of stellar objects form a discernible pattern; usually named after mythological gods, animals, objects, or people.

The stars visible in the night sky have always fascinated humans. Ancient humans were very dependent on stars, assigning a mystic aspect to them that influenced political and social decisions of the period. Names were assigned to the stellar patterns to produce a sense of order and purpose to the cosmos. The majority of the constellation names are based on Greek Mythology. Examples include Orion, Taurus and Pegasus. However, because the patterns appear slightly different depending on the viewer's geographical location, different names are assigned to the constellations by different cultures. For example, in the Chinese culture Taurus the Bull is the Snake and Aquarius is the Tiger.

The ancient observers believed that the Sun changed places with the stars to create nighttime. These visible stars formed the basis of the **Zodiac Constellations**: the first constellations to be named by early astronomers. The scholars who studied the Zodiac Constellations were **Astrologers**: people who studied the stellar objects in the Zodiac Constellations in order to predict the future. Besides the original twelve constellations comprising the Zodiac Constellation, modern astronomy has found and named an additional 76 others for a total of 88 constellations. Consequently, this divides the sky into 88 sections.

### Locating the Stars

The positions of the stars are referenced in relation to the Earth and are described in terms of the celestial coordinates of Right Ascension and Declination.

The declination and right ascension is described in reference to the **Celestial Sphere/Globe Model.** You can picture this model as having the Earth centered in the middle of a sphere. The outer framework of the sphere rotates clockwise (east to west) in relation to the Earth. The north and south poles of the Earth correspond to the north and south celestial poles. The Earth's equator corresponds to the celestial equator. **Right Ascension** is roughly analogous to lines of Longitude and is measured eastward from the vernal equinox. **Declination** is roughly analogous to lines of Latitude and is measured in relation to the celestial equator: positive to the north, and negative to the south.

Right Ascension and Declination are measured in units of degrees and time. 1 sec = 1/3600 of a degree. Example: 15° 12' 5" Right Ascension. The sky shifts 15° every hour. That's why photo-telescopes must move with the stars.

**Celestial Brightness**

Over 2,000 years ago the Greek astronomer Hipparchus devised a system to classify the brightness of the stars. He ranked the brightest stars as first-class stars, the next brightest as second-class, etc., down to the faintest stars, the sixth-class. Although this system worked well for millennia, with the advent of more powerful telescopes, astronomers discovered stars that were both brighter and fainter than the brightest and faintest stars in the star catalog. To solve this problem modern astronomers extended the magnitude system into negative numbers to account for the brighter stars. Example: Sirius, the brightest star in the sky, has a magnitude of -1.42.

These numbers represent a star's **Apparent Visual Magnitude ($m_v$)**: a star's brightness as seen from Earth. On the magnitude scale, the larger the number, the fainter the star.

The relative brightness of the stars in a particular constellation is denoted by the use of lower case Greek letters; the brightest appearing star in the constellation being Alpha ($\alpha$) and the faintest appearing star being Omega ($\omega$). The Greek alphabet symbols are used as a prefix appended to the constellation's name. The possessive form of the name is used, and this is created by appending the suffix 'is' to the name of the constellation. Example: The brightest star in the constellation Orion is noted as Alpha Orionis.

However, stellar brightness requires a more precise measurement system beyond the original model because variances in the human eye and weather conditions can dramatically affect the apparent magnitude of a star.

For the purposes of mathematical measurement, astronomers use a star's **Intensity:** a measure of the light energy from a star that hits 1 square meter in 1 second.

We can use this modern system in conjunction with the Hipparchus devised brightness classifications of first-class, etc., because a difference of 5 magnitudes corresponds to an intensity ratio of 100. This allows us to compare the intensity of one star to another based on the mathematical progression of the difference in magnitude. This progression is expressed in the formula
$I_A/I_B = (2.512)(m_B - m_A)$.

**SKILL 11.3**     **Describe the evidence for and interpretations of an expanding universe (e.g., red shift and the Doppler effect).**

**Detecting Motion of Stars**

**Doppler Effect**: the apparent change in frequency of light or sound that occurs when the source of the wave is moving relative to the observer. The Doppler effect can be used to measure the motion of an object both on Earth and in space.

The motion of the stars can be determined by observing the Doppler effect associated with their movements. As a star moves away from us it has a **Red Shift** and its emitted light wavelength is longer. As a star moves toward us, it has a **Blue Shift** and its emitted light wavelength is shorter. Red and Blue shift doesn't refer to the color of the star. It refers to the movement of the $\lambda$ along the spectrum. By comparing the $\lambda$ of a star at different times, we are able to determine its shift. We find the $\lambda$ shift by examining the object's spectral properties. Doppler can also detect rotation. We discover differing wavelengths because one side is closer and the other side is farther away.

**SKILL 11.4**     **Analyze types of evidence used to infer scales and relative motions of the solar system, the Milky Way galaxy, the universe (e.g., in relation to relative size and distance).**

The solar system is divided into two sections: the inner and outer planets. The inner planets' composition reflects the attraction of the heavier elements by the Sun. The outer planets' composition reflects the lighter, less dense elements not attracted as much by the Sun's gravitational mass. Heavy elements sink inward to form the core. Lighter elements form the atmosphere.

Our solar system consists of the Sun, planets, comets, meteors, and asteroids. The planets are ordered in the following way: Mercury, Venus, Earth, Mars (with Asteroid Belt), Jupiter, Saturn, Uranus, and Neptune. Pluto has historically been considered the ninth planet. Recently, scientists have created new requirements for the definition of a planet, and it is possible that Pluto will be considered only a celestial body. A common memory aid for remembering the order of the planets is <u>M</u>y <u>V</u>ery <u>E</u>ducated <u>M</u>other <u>J</u>ust <u>S</u>erved <u>U</u>s <u>N</u>ine <u>P</u>ies.

All of the planets revolve around the Sun, and all of the planets (with the exception of Venus) rotate on their axis in the same direction. Venus has a retrograde motion; it rotates backwards. Except for Pluto, all the planets follow roughly the same elliptical orbital planes around the sun. Neptune and Pluto occasionally change places in the order. Pluto's orbit is very erratic compared to the other planets and sometimes it carriers Pluto inside of Neptune's orbit. The asteroid belt is located between Mars and Jupiter and may be the remnants of a planet crushed by the massive gravitational force of Jupiter.

| Comparison of the Basic Characteristics of the Inner and Outer Planets: | |
|---|---|
| INNER PLANETS | OUTER PLANETS |
| Referred to as the Terrestrial Planets | Called the Gas Giants or Jovian Planets |
| Similar to density to Earth. | Except for Pluto, very large in size. |
| Also referred to as the "Rocky Planets." | Primarily composed of gas. |
| Relatively small in size. | Less dense than Earth. |
| Spin slowly on their axis. | Rotate rapidly on their axis. |
| Few if any, moons. | Lots of moons. |
| Mercury | Jupiter |
| Venus | Saturn (Ringed) |
| Earth | Uranus |
| Mars | Neptune |
| + Asteroid Belt | Pluto (Rocky) |

**Measurement Units in Astronomy**

Astronomical distances represent mind-boggling amounts of distance. Because our standard units of distance measurement (i.e. kilometers) would result in so large of a number as to become almost incomprehensible, physicists use different units of measurement to reference the vast distances involved in astronomy.

Within our solar system, the standard unit of distance measurement is the **AU (Astronomical Unit)**. The **AU** is the mean distance between the Sun and the Earth. 1 AU = $1.495979 \times 10^{11}$ m. Outside of our solar system, the standard unit of distance measurement is the **Parsec**. 1 Parsec = 206,265 AU or 3.26 Light Years.

**Light year (LY):** the distance light travels in one year. The speed of light is $3.00 \times 10^{8}$ m/sec, so one light year represents a distance of $9.5 \times 10^{12}$ km, or 63,000 AU.

**EARTH SCIENCE**

**Measuring the Distance to the Stars**

The distances are measured using a shift in viewpoint. This is called the **Parallax**: an apparent change in the position of an object due to a change in the location of the observer. In astronomy, parallax is measured in seconds of arc, where 1 second of arc = 1/3600 of a degree.

The concept of measuring distances by parallax is based on the mathematical discipline of trigonometry. Example: Photos of the distant stars are taken at different times, usually 6 months apart. The apparent shift in position that the star has moved in comparison to the previous photo is the parallax.

By measuring the angles, we can use the trigonometric functions of sine, cosine, and tangent to determine a distance to the object. The smaller the parallax is, the greater the distance to the star.

| SUBAREA III. | METEOROLOGY |
|---|---|

**COMPETENCY 12.0**     Understand the composition, structure, and properties of the Earth's atmosphere and the mechanisms and effects of energy transfer involving the Earth-atmosphere system.

**SKILL 12.1**     Identify the properties (e.g., density, composition, temperature) of the atmosphere from the Earth's surface through the thermosphere and the significance of changes in these properties.

**Composition of the Atmosphere**

The Earth's atmosphere is very similar to a fluid. The atmosphere makes up only 0.25% of what we call the Earth, and like the fluids of the ocean, our atmosphere is driven by heat, primarily solar radiation. Having an atmosphere is not unique for a planet. To a degree, most of the planets in our solar system have an atmosphere. However, the presence of significant oxygen in Earth's atmosphere is unique and makes life possible on our planet. Earth's atmosphere is composed of 78% Nitrogen, 21% Oxygen, and 1% other gasses.

**Components of the Atmosphere**

Water Vapor: Along with Carbon Dioxide ($CO_2$) and Methane, Water Vapor ($H_2O$) is considered a **Greenhouse** Gas: a gas that absorbs heat energy. Water Vapor is the most prevalent of the greenhouse gasses and is especially good at collecting heat energy, as evidenced by the Earth's ability to retain heat at night when solar radiation is lowest.

Dust and Aerosols: These are natural components of the atmosphere and their presence produces optical phenomena such as making the sky appear blue, rainbows, and the Northern and Southern Lights.

Pollutants: Some are man-made, including industrial waste, chemical refrigerants and hydrocarbons released from the burning of fossil fuels. Some are natural such as Terpene, which is released from trees, Saharan Dust, and $CO_2$ released by volcanoes.

**Layers of the Atmosphere**

**Troposphere**: (Ground level to 11Km, 0 to 17.6 miles, or 0 to 92,928 feet): The Troposphere varies in height according to the temperature. It is lower at the poles and higher at the equator. Because the pressure decreases, it gets colder as you go up in the Troposphere. Only very rarely do you have a mixing between the Troposphere and the next layer, the Stratosphere. All storms, weather fronts, and weather occur in the Troposphere.

**Stratosphere**: (11Km to 50Km, 17.6 to 80 miles, or 92,980 to 422,400 feet): The Stratosphere is characterized by weak vertical air motion and strong horizontal air motion. There is very little lifting or sinking air in the Stratosphere. Temperatures warm as you go up due to the presence of the Ozone layer contained within the Stratosphere.

**Mesosphere**: (50 to 85Km, 80 to 136 miles, or 422,400 to 718,080 feet): It is bitterly cold in the Mesosphere.

**Thermosphere**: (85 to 600Km, 136 to 960 miles, or 718,000 to 5,068,800 feet): This is the hottest portion of the atmosphere with rapid warming accompanying a rise in altitude. There are very few molecules left to block out the incoming solar radiation. The outer reaches of the Thermosphere are also sometimes referred to as the Exosphere.

**Ionosphere**: (Located within the upper portion of the Mesosphere at 80Km and goes into the Thermosphere). The ionosphere is an area of **free ions**: positively charged ions, produced as a result of solar radiation striking the atmosphere. The solar wind strikes the Ionosphere at the polar dips in the Magnetosphere. The ions are excited to a higher energy state and this energy is released into the visible spectrum to form the Aurora Borealis (Northern Lights).The Ionosphere varies with the time of day, season, and Sunspot cycles. When the Sun sets at night, less ions strike, extending radio wave communications. There is more radiation during Sunspot cycles. These hypercharge the atmosphere and can disrupt radio waves during the daytime.

Ozone Layer ($O_3$): (Contained within the Stratosphere): Ozone is essential to life on Earth and is continually formed and destroyed within the atmosphere. Only a very thin layer of ozone protects against UV (ultra violet) radiation. Ultra violet radiation scrambles the DNA codes in human cells, and can kill the cells or, at a minimum, cause cancer.

Much concern exists about a hole in the ozone layer. This is a misnomer. In reality, there is not a hole, but a possible thinning of the layer. Many scientists believe this is due to the presence of CFC's (Carbon Fluorocarbons). The chlorine (Cl) in CFCs, as well as from other sources, steals an oxygen atom from Ozone ($O_3$) molecules. This theft leaves behind only plain oxygen ($O_2$), which does not effectively screen out UV radiation. However, the resultant ClO molecule is very unstable and UV radiation can easily break it apart. The released chlorine then attacks another ozone molecule and the process repeats itself.

CFCs were thought to be the primary culprit, but this theory is questioned. CFCs were only invented in the 1920's for use in aerosol spray cans, industrial processes, and refrigerants. Although it may be a contributor in the depletion of the ozone layer, not all scientists believe the theory that CFC's are solely responsible. Why does the thinning occur over only the Antarctica? The greatest use of CFCs occurred in the industrialized nations of the Northern Hemisphere. Additionally, the hole varies in size from year to year, appearing during the Antarctic spring in October and disappearing by mid November or December.

There is a lack of data. Data has been collected on this thinning only since 1979. There is a possibility that the hole may have been there before the introduction of CFCs. An alternate theory to CFCs is **Circumpolar Vortex**.

Because there is a great deal of Open Ocean at Antarctica, it is a very cold place during the winter. The cold, in effect, isolates the Antarctic atmosphere from the rest of the warmer atmosphere. This extreme cold forms ice crystals in the atmosphere and Chlorine (Cl) is locked into the crystals. When the Antarctic spring arrives, the atmosphere thaws out, releasing the chlorine into the atmosphere. The chlorine attacks the ozone layer, decreasing its density. As spring progresses, the circumpolar vortex weakens, allowing the air to mix with the normally ozone rich air. Thus, the hole disappears.

**SKILL 12.2**      **Analyze how various wavelengths of solar radiation (e.g., ultraviolet, visible light, infrared) are affected as the radiation enters and passes through the atmosphere and is absorbed by and radiated from the Earth's surface.**

The Sun emits light at different wavelengths in the electromagnetic spectrum. UV light is electromagnetic radiation with a wavelength shorter than that of visible light. The three types of UV light (near UV, far UV, and deep UV) fall within the wavelength range of 200 to 1 nm.

Visible light is the portion of the electromagnetic spectrum that can be detected by the human eye. This type of electromagnetic radiation typically falls within the wavelength range of 400 to 700 nm, and is also known as short wave radiation.

Infrared (IR) radiation is electromagnetic radiation of a wavelength longer than that of visible light. Infrared radiation falls within the wavelength range of 750 nm to 1 mm, and is also known as long wave radiation.

The Earth's atmosphere selectively controls the passage of solar radiation towards the Earth's surface. Thus, not all solar radiation received at the periphery of the atmosphere reaches the Earth's surface. Upon entering the Earth's atmosphere, large amounts of solar radiation can be absorbed or scattered in all directions by atmospheric gases, vapors, and dust particles.

The scattering of solar radiation is the process by which small particles and gas molecules diffuse part of the incoming solar radiation in random directions without altering the wavelength of the electromagnetic energy. There are two types of atmospheric scattering: selective scattering and non-selective scattering. These two processes are determined by the different sizes of particles in the atmosphere.

Selective Scattering

Selective scattering occurs when atmospheric gases or particles are smaller in dimension than the wavelength of a particular type of solar radiation. During the process of selective scattering, radiation with shorter wavelengths are scattered more extensively than those with longer wavelengths. Thus, selective atmospheric scattering is inversely proportional to the wavelength of solar radiation and has an increasingly large effect on distinct types of light in the following order: Infrared Light < Visible Light < UV Light. The scattering effect on UV light is approximately ten times as great as on the red rays of sunlight.

Non-Selective Scattering

This type of scattering occurs in the lower atmosphere due to the presence of dust, fog, and clouds with particle sizes more than ten times the wavelength of the components of solar radiation. The amount of non-selective scattering is equal for all wavelengths.

Absorption

Atmospheric absorption is the process in which solar radiation is retained by a substance, such as an atmospheric particle or gas, transferred into heat energy and then converted into long wave radiation emissions. The degree to which a certain type of solar radiation is absorbed is wavelength-dependant. Most entering ultraviolet radiation measuring less than 200 nm in wavelength, is absorbed by oxygen and nitrogen in the ozone layer of the upper atmosphere. Infrared solar radiation, with its wavelengths greater than 700 nm, is absorbed to a smaller extent by vaporous or condensed atmospheric carbon dioxide, ozone, and water.

Due to the absorption and scattering effects of the atmosphere, little UV radiation and only 50% of short wave radiation actually interacts with the Earth's surface. When short wave radiation reaches the surface, it is absorbed, transformed into heat, and re-emitted in the form of long wave radiation (infrared radiation).

**SKILL 12.3** **Identify the processes by which energy in transferred to and within the atmosphere (e.g., radiation, convection, conduction).**

Heat is transferred in three ways: **conduction, convection, and radiation.**

**Conduction** occurs when heat travels through the heated solid.

The transfer rate is the ratio of the amount of heat per amount of time it takes to transfer heat from an object to another area. For example, if you place an iron pan on a flame, the handle will eventually become hot. How fast the handle gets too hot to touch is a function of the amount of heat and how long it is applied. Because the change in time is in the denominator of the function, the shorter the amount of time it takes to heat the handle, the greater the transfer rate.

**Convection** is heat transported by the movement of a heated substance. Warmed air rising from a heat source, such as a fire or electric heater, is a common example of convection. Convection ovens make use of circulating air to more efficiently cook food.

**Radiation** is heat transfer as the result of electromagnetic waves. The Sun warms the Earth by emitting radiant energy.

An example of all three methods of heat transfer occurs in the thermos bottle or Dewar flask. The bottle is constructed of double walls of Pyrex glass that have a space in between. Air is evacuated from the space between the walls and the inner wall is coated in silver. The lack of air between the walls decreases heat loss by convection and conduction. The heat inside is reflected by the silver, cutting down heat transfer by radiation. Hot liquids remain hotter and cold liquids remain colder longer.

**SKILL 12.4    Analyze global wind patterns in terms of latitudinal variations in insolation and the Corilois effect.**

**Solar Radiation (Insolation)**

This distribution of solar energy is called **insolation**. Solar radiation isn't distributed evenly across the Earth because of the Earth's curvature, axial tilt, and orbit. This results in uneven heating of the atmosphere, and is why the temperature is warmer at the equator and colder at the poles.

Because of Earth's curvature and tilt, the energy striking the polar areas is spread over a larger area. At the equator it is more concentrated. The same amount of energy is striking the atmosphere, but it is striking a larger or smaller area. In effect, this dilutes the energy received by a particular area.

The effect of insolation is very important to life on Earth. The absence of solar radiation would cause the creation of very cold air masses and the thermal blanket of the atmosphere would not have heat to hold and reradiate. In short order, the world would become an icy rock.

**Air as an Insulator**

Air in the atmosphere is a good insulator, analogous to a goose-down blanket.

**Convection Cells**: The ground absorbs energy during the morning hours and re-radiates the absorbed energy in a narrow band near the surface. This heats the air near the surface. The air near the surface absorbs the energy and expands. Convection is the primary means of transferring heat to the air.

As the molecules move faster (because of the heat transfer), the air expands, becomes less dense, and rises. The air continues to rise, but begins to cool, again becoming denser. Then it starts to sink back to the surface. This cycle of rising, warming air and sinking, cooling air is called a convection cell.

**Parcel of Air (Air Mass):** a large area of air, which assumes a characteristic temperature, pressure, and humidity from sitting over a landmass. Horizontally, the parcel has relatively uniform temperature, pressure, and humidity. Vertically, the parcel may have widely differing temperatures, pressures, and humidity. *Remember: when you think of air parcels, think in the horizontal plane.*

## Coriolis Effect

The **Coriolis Effect** is the deflection of air or water currents caused by the rotation of the Earth. This creates global wind patterns that affect the climate. These wind patterns also represent rain patterns. The wind patterns between 30°N and 0°N are called the **Trade Winds**. The wind patterns between 30°N and 60°N are called the **Prevailing Westerlies**. All the great deserts of the world lie between the Tropics at 0° and 30°North and South latitudes. A shift in the wind patterns would also shift the deserts. The optimum growing zone is between 30° and 60° North and South latitudes.

If global warming takes place, it could cause the melting of the polar ice caps that in turn could raise sea levels. This would inundate land areas and lead to economic disaster.

The Earth's climate is in a very delicate balance. A change to the global warming or cooling patterns would affect retained heat, which would further affect the growing zones. In example, global warming would cause the expansion of the tropics, while global cooling would cause a contraction of the tropic zone.

## Global Wind Patterns

The rise and fall of heat at the 0°, 30°, 60° and 90° latitudes drive convection cells by causing the pressure gradients to speed up or slow down. The **Jet Streams** are zones of very strong, moving air confined to narrow columns and marks the zones where the cold Polar air and the warmer air meet. This produces the greatest pressure gradients. The Jet Streams can be either straight or dramatically dip, thus creating ridges and troughs on the 500-300mb pressure surface.

The flatter the isobars, the more evenly balanced the weather. The more pronounced the ridges and troughs, the more pronounced the swings in the weather.

The ITCZ (**Inter Tropical Convergence Zone**) controls the weather in the tropics, and it moves north and south of the equator. The ITCZ is responsible for the formation of Monsoon rains.

The **Horse Latitudes** are located between 0° and 30° north and south latitudes.

The **Doldrums**, an area of no wind, is located at 0°.

The **Trade Winds** are very strong and blow all the time in the Horse Latitudes. The Trade Winds also provide direct heating to the coastal climate in this zone.

The **Prevailing Westerlies** are found between 30° and 60° north and south latitude. These cause storms and winds to move in a west to east pattern.

**Polar Winds** are the product of the presence or absence of sunlight, not polar cells.

### Equation of Motion (EOM)

The Equation of Motion predicts how water moves. Wind is the initial mover of water. The moving surface water couples to water below it and moves the deeper water in varying degrees according to depth.

The general equation for dealing with force is F= Ma, where Force = mass times acceleration). However, for oceanography, the terms used are somewhat different and more specialized to the field.

**Pressure Gradient** = gi where g=gravity and i=the slope of the pressure gradient

**Coriolis Effect** =fV where f=planetary vorticity, and V=velocity.

**Wind Stress** = A $d^2$V where A=area, $d^2$ =diameter squared, and V=velocity.

**Pressure Gradient**: the difference between high and low pressure areas. The difference between the areas represents a range of varying high and low pressure. Changes in temperature and density occur within this area. The pressure gradient is a major player in determining ocean circulation. Example: The Gulf Stream is one of the strongest, most consistent ocean currents.

**COMPETENCY 13.0**      Understand the properties of water, conditions in the atmosphere that favor phase changes, and the energy relationships among phase changes, cloud formation, and precipitation.

**SKILL 13.1**      Relate the physical properties of water (e.g., high specific heat, surface tension) to the chemical structure and properties of water molecules.

**Significance of Water**

Water ($H_2O$) is significantly different from its immediate Hydrogen compound cousins. **Compounds** are substances that contain two or more elements in a fixed proportion. Generally, the heavier molecules have higher boiling and freezing states based upon molecular weight.

A group of atoms held together by chemical bonds is called a **molecule**. The bonds form when the small, negatively charged electrons found near the outside of an atom are shared or transferred between the atoms. The bonds formed by the shared pair of electrons are known as **covalent bonds**.

Most substances tend to adopt a solid or gaseous form. Water is different. It wants to be a liquid. A water molecule forms when covalent bonds are established between two hydrogen atoms and one oxygen atom.

However, unlike the other hydrogen compounds, water has its two hydrogen molecules on one side of the atom. It's a polar molecule.

This arrangement of atoms is based on the distribution of the water molecule's oxygen electrons. The electrons cause the geometric shape of the molecule to be angular. This angular shape makes the molecule electrically asymmetrical (polar).

This polar arrangement gives water some very special properties.

- The polar molecule acts similar to a magnet. Its positive ends (the hydrogen) attract particles having a negative charge, and its negative end (the oxygen) attracts particles that have a positive charge. This arrangement is the basis for the water molecule's rightful description as the "**Universal Solvent**." When water comes into contact with compounds- for example salts- held together by the attraction of opposite charges, the water molecule separates that compound's elements from each other.

- Another unique property of water is that water likes itself; it has a natural tendency to stick to itself. Once again, this property is based upon the polar nature of the water molecule. It attracts other water molecules. When the molecules stick together, they are attached through Hydrogen Bonds, giving the molecule a property called **cohesion**. Cohesion gives water an unusually strong surface tension, and its capillary action makes the water spread. When the water spreads, **adhesion**, the tendency of water to stick to other materials, allows water to adhere to solids, making them wet.

- Water is the only known substance that readily exists in all three states of matter: liquid, solid, and gas. The hydrogen bonds holding water together are important. If they didn't exist, water would fly apart to form a gas. However, water primarily wants to be a liquid, and its liquid state range is averaged at 16°C.

**Water and the States of Energy**

**Specific Heat**: the amount of heat required to raise or lower the temperature of 1 gram of a substance by one degree Celsius.

**Heat Capacity**: the ability of a substance to resist a change in temperature. Water has an extremely high heat capacity. It is very resistant to changes in temperature.

It takes a great deal of energy transfer to heat up water, but it also takes a great amount of energy loss to cool it down. When we heat water we are actually transferring energy from a heat source to the water. The water molecules absorb the energy and move faster. This causes the temperature of the water to rise.

Matter can change state if enough heat (energy) is applied or removed.

**SKILL 13.2** **Describe the energy changes involved in the transition between phases of water (i.e., latent heat).**

**Heat of Fusion**: the amount of heat energy expressed in calories required to change 1 gram of water from a solid to a liquid, or back. The heat of fusion is equal to 80 cal/gram.

There is a difference in the amount of heat required to change states between liquids, solids, and vapors. It takes more heat to change from a liquid to a vapor, or a vapor to a liquid, than it does to change from a solid to a liquid, or a liquid to a solid.

Water is a dynamic factor in our atmosphere because its resistance to change state allows the Earth to maintain a temperature balance in the presence and absence of solar radiation (i.e. day and night).

**Latent Heat**: the amount of heat required to change matter from one state to another. Example: The amount of heat required to cause evaporation or condensation of water.

**Heat of Vaporization**: the amount of heat energy required to change 1 gram of water from a liquid to a vapor, or back. The heat of vaporization is equal to 540 cal/gram. Example: 1 gram of water at 99 degrees Celsius requires 1calorie of heat to raise the temperature to 100°C, which is the boiling point of water. However, the water will not turn into steam (vapor or gas), until an additional 540 calories are added because of the high heat capacity of water. The requirement for additional calories to force a change of state works in our favor. Without the law of heat of vaporization, the oceans would rapidly vaporize!

Rain represents the opposite pattern. In rain, 540 calories have been removed, changing the state of water from a vapor (gas) to a liquid.

**SKILL 13.3    Describe the atmospheric conditions under which fog and clouds with various characteristics form (e.g., adiabatic temperature changes, dew-point, atmospheric stability).**

**Fog**: a cloud that touches the ground. Fog forms when cold air moves over a warmer surface. Fog is very common along shorelines because the specific heat of water retains heat and is consequently much warmer than the overlaying air. Fog can also form inland where the same basic conditions exist.

### Clouds

Clouds are classified by their physical appearance and given special Latin names corresponding to the cloud's appearance and the altitude where they occur. Classification by appearance results in three simple categories: cirrus, stratus, and cumulus clouds. Cirrus clouds appear fibrous. Stratus clouds appear layered. Cumulus clouds appear as heaps or puffs, similar to cotton balls in a pile. Classification by altitude results in four groupings: high, middle, low, and clouds that show vertical development. Other adjectives are added to the names of the clouds to show specific characteristics.

### Calculating Dew Point

Calculating dew point is based upon a comparison of the difference between the dry bulb temperature versus the wet bulb temperature. Standardized tables are used to predict the dew point. This is calculated by taking the dry bulb temperature and subtracting the wet bulb temperature. The difference is then indexed against the dry bulb temperature.

## Cloud Classifications

**High Clouds**: -13 °F (-25 °C): Composed almost exclusively of ice crystals.
    Cirrus >23,000 ft (7,000 m). Nearly transparent, delicate silky strands (mare's tails) or patches.
    Cirrostratus >23,000 ft (7,000 m). A thin veil or sheet that partially or totally covers the sky. Nearly transparent, the sun or moon readily shines through.
    Cirrocumulus >23,000 ft (7,000 m). Small, white, rounded patches arranged in a wave or spotted mackerel pattern.

**Middle Clouds**: 32 to -13 °F (0 to -25 °C): Composed of super-cooled water droplets or a mixture of droplets and ice crystals.
    Altostratus 6600 - 23,000 ft (2000 - 7000 m). Uniform white or bluish-gray layers that partially or totally obscure the sky layer.
    Altocumulus 6600 - 23,000 ft (2000 - 7000 m). Roll-like puffs or patches that form into parallel bands or waves.

**Low Clouds**: > 23 °F (-5 °C): Composed mostly of water droplets.
    Stratocumulus 0-6,600 ft (0-2000 m). Large irregularly shaped puffs or rolls separated by bands of clear sky.
    Stratus 0-6600 ft (0-2000 m). Uniform gray layer that stretches from horizon to horizon. Drizzle may fall from the cloud.
    Nimbostratus 0-13,120 ft (0-4000 m). Thick, uniform gray layer from which precipitation (significant rain or snow) is falling.

**Clouds with Vertical Development**: Water droplets build upward and spread laterally.
    Cumulus 0-9840 ft (0-3000 m). Resemble cotton balls dotting the sky.
    Cumulonimbus 0-9840 ft (0-3000 m). Often associated with thunderstorms, these large puffy, clouds have smooth or flattened tops, and can produce heavy rain and thunder.

## Expansion, Compression, and the Adiabatic Process

### Expansional Cooling and Compressional Warming

One way that air cools and increases in relative humidity is by **Expansional Cooling**. As a parcel of air is heated by the Earth's surface, it will start to rise. This rising air encounters lower pressures as it increases in altitude, and will start to expand. As the air expands in volume, it begins to cool. With the expansion in volume, the molecules don't strike each other as often, and without this release of kinetic energy, the temperature goes down. Example: This same principle applies when you let air out of or put air into an automobile tire. Feel the outgoing air from a tire. It is cool. That's because as the pressure is decreased the air expands and the temperature drops.

**Compressional Warming:** utilizes the reverse principle. When you place air under pressure, it occupies less volume. This in turn causes the molecules to strike each other more often. Consequently, the energy released cause a rise in temperature. In the atmosphere, compressional warming causes a decrease in relative humidity.

**The Adiabatic Process**

**Adiabatic Process**: a process in which heat is neither absorbed nor taken away. The Expansional Cooling and Compressional Warming of unsaturated air in the atmosphere are adiabatic processes. During its ascent or descent, a parcel of air is neither heated nor cooled by radiation, conduction, phase changes of water, or mixing with its surroundings. Heating and cooling is solely accomplished by the expansion or contraction of the air in response to pressure changes.

The rate of cooling and warming is expressed as the **Dry Adiabatic Lapse Rate** (lapse refers to temperature change with altitude).

- Rising air cools: The cooling rate is 10.0 °C per 1000-meter increase in altitude (5.4 °F per 1000 feet increase in altitude).
- Sinking air warms: The warming rate is 10.0 °C per 1000-meter decrease in altitude (5.4 °F per 1000 feet decrease in altitude).

However, when air cools to the point of reaching saturation (100% humidity), the effect of expansional cooling is partially negated, and the Wet Adiabatic Lapse Rate is used to express the cooling rate. Unlike the dry lapse rate, the wet lapse rate varies with temperature. This in turn means that some latent heat is released which offsets the expansional cooling effects.

For convenience, the average value of 6°C per 1000-meter increase in altitude, 3.3°F per 1000 feet increase in altitude, is used as the **standard Moist Adiabatic Lapse Rate**.

**Atmospheric Stability**

Warm air rises and will continue to rise until it meets air that is as cold or colder than the rising air. When rising air cools to the dew point, clouds form due to condensation. Air can rise even if it is only a few degrees warmer than the surrounding air.

Atmospheric stability is determined by comparing the temperature change of an ascending or descending air parcel with the temperature profile of the ambient air layer in which the parcel ascends or descends. Air is considered as either **stable** or **unstable**.

Air layers become more stable when they descend, and less stable when they ascend. Unstable air will remain unstable until it encounters air of the same temperature or cooler.

In stable air, there is no vertical movement. In unstable air, there is a great deal of vertical movement.

**Stable Air**

Stable air exists when there is no vertical movement of the air. An ascending air parcel becomes cooler (denser) than the ambient air, and a descending air parcel becomes warmer (less dense) than the ambient air. Any upward or downward displacement of an air parcel in stable air gives rise to forces that tend to return the parcel to its original altitude.

Once stable air becomes colder than the surrounding air it sinks. It will sink to the altitude in which the temperature of the ambient air equals that of the sinking air. If a warm air mass is over a cold air mass, the air is very stable.

During the nighttime, **Radiational Cooling** of the ground tends to stabilize the overlying air. However, during the daytime, solar heating of the ground tends to destabilize the overlying air masses.

**Unstable Air**

Stable air can easily become unstable through **Radiational Heating**. As the ground heats up, the air also heats up and becomes unstable. The unstable air will continue to rise vertically until it encounters air as cold, or colder, than it is.

The stability of air is affected by the relative humidity of the air mass, which can cause **Conditional Stability**. The air mass is considered stable if the environment is dry, but is unstable if the air becomes saturated.

Stability is an ever-changing situation throughout the day due to heating of the Earth's surface and atmosphere. Sometimes a layer of stable air can become trapped between two layers of unstable air. This is called an **inversion** and is the result of the fact that stability can change with altitude.

Air tends to stabilize over a colder surface and destabilize over a warmer surface.

**SKILL 13.4**     **Describe the conditions under which precipitation forms.**

**Causes of Precipitation**

**Collision & Coalescence**: In warm clouds the water vapor moves, merges, and collides to form bigger drops. Generally, the darker a cloud is, the bigger the drops formed. Eventually, the drops get too heavy for the supporting updrafts and drop out of the clouds as rain.

**Bergeron Process**: In a cold cloud (below freezing), ice begins to form into crystals and attracts other water droplets. Eventually the crystals get too heavy and drop out of the cloud.

**Forms of Precipitation**

All forms of precipitation start from an interaction of water vapor and other particulate matter in the atmosphere. These particulates act as a nucleus for raindrops, as the water vapor particles attach themselves to the other airborne particles.

Because one of water's major properties is that its water particles attract other water particles, the raindrop grows as water vapor particles accrete around the nuclei.

**Drizzle**: any form of liquid precipitation where the drops are less than 0.02 inches in diameter.

**Rain:** any form of liquid precipitation where the drops are greater than 0.02 inches in diameter.

**Virga**: the meteorological condition where rain evaporates before touching the ground. You see it rain, but it never hits the ground.

**Snow**: water molecules that form into ice crystals through freezing. The shape of the snowflakes depend on the temperature at which they formed:

Needles = 0°C to -10°C.
Dendrites= -10°C to -20°C.
Plates= -20°C to -30°C.
Columns= -30°C to -40°C.

**Freezing Rain**: drops fall as rain but immediately freeze upon hitting a cold surface such as power lines, roofs, or the ground. This is also called an **Ice Storm**.

**Rime Ice**: ice droplets that have tiny air bubbles trapped within the ice, producing an opaque layer of granular ice.

**Sleet**: officially called ice pellets, these are drops of rain 5mm or less in diameter. Sleet freezes before hitting the ground and bounces when it strikes a surface.

**Hail**: precipitation in the form of balls or lumps of ice. Hail forms when an ice pellet is transported through a cloud that contains varied concentrations of super-cooled water droplets. The pellet may descend slowly through the entire cloud, or it may be caught in a cycle of updraft and downdraft. The ice pellet grows by accreting (adding) freezing water droplets. Eventually the weight of the hail grows too heavy to be supported by the air column and falls to the ground as a hailstone. The size of the stone depends on the amount of time spent in the cloud.

**Fog**: a cloud that touches the ground. Fog forms when cold air moves over a warmer surface. Fog is very common along shorelines because the specific heat of water retains heat and is consequently much warmer than the overlaying air. Fog can also form inland where the same basic conditions exist.

**Dew Point**: the temperature to which air must be cooled for condensation to occur. If there is a large difference between the dew point and the temperature, then you have low humidity. If there is a small difference between the dew point and the temperature, then you have a higher humidity. Remember: Colder air has a lower water vapor capacity than does warm air. That is why you can have high humidity in cold air and medium humidity in warm air. Ironically, the warm air actually has more water vapor present (specific humidity).

**SKILL 13.5** **Predict the type of precipitation that will fall to the Earth's surface under given conditions.**

### Rain

Warm air turns the water from rivers, lakes, and oceans into water vapor. As warm, moist air rises, it begins to cool, causing water vapor to condense and form clouds. Precipitation is generated by cumulonimbus clouds, which typically appear dark and gray, and can produce lightening and thunder. Once water droplets of cumulonimbus clouds have grown large enough, two processes can occur to form rain: Coalescence and the Bergeron process.

## Coalescence

Coalescence occurs when water droplets fuse to create larger water droplets. Air resistance keeps the water droplets in a cloud stationary, but when air turbulence occurs, water droplets collide, producing larger droplets. As these larger water droplets descend, coalescence continues, and the drops eventually become heavy enough to overcome air resistance and fall to the Earth's surface as rain (if cloud, air, and surface temperatures are above 32° F).

## Bergeron process

Rain can also begin as ice crystals that collect to form snowflakes in higher cumulonimbus clouds. The Bergeron process occurs when ice crystals acquire water molecules from nearby super-cooled water droplets and begin to fall. This process is temperature dependent, as super-cooled water droplets only exist in clouds below freezing. When ground temperature is above 32 degrees F, falling ice crystals melt as they pass through the freezing level into warmer air and reach the surface as rain.

## Snow

Snow occurs when the layer of the atmosphere from the surface of the Earth through the cloud is entirely below freezing. In this case, precipitation forms through the before-mentioned processes and falls from the cloud as snow. Because air and surface temperatures are below freezing, the falling snow does not melt and reaches the Earth as ice crystals.

## Hail

Hail are large, frozen raindrops produced by intense thunderstorms. This condition is rarely seen when the surface temperature is below freezing. Hail is formed as a byproduct of strong updrafts that exist in thunderstorms. The cumulonimbus clouds that are associated with thunderstorms can grow to heights where the temperature is below freezing. Drops of water that rise with the upward directed winds associated with thunderstorms collide with other droplets and continue to rise and cool, eventually creating a large hailstone. This hailstone will continue to grow as it cycles through the cloud, rising and falling, until it grows too large for updrafts. At this point, the hailstone will fall to the ground. The stronger the thunderstorm, the stronger the updraft, and the longer the hailstone will grow in size.

## Freezing Rain

Freezing rain occurs when there is a large layer of warm air above a shallow layer of below freezing air at the Earth's surface. Precipitation that falls from the cloud through the warm air reaches the shallow layer of cold air as rain. When it hits the sub-freezing layer its precipitation cools. These super-cooled water droplets then freeze on contact as they reach the Earth's below freezing surface.

## Sleet

Sleet occurs when there is a small layer of warm air above a very deep layer of below freezing air at the surface. In this case, precipitation falling through the warm air will reach the layer of cold air as rain. As these raindrops enter the cold air, they freeze and reach the ground as frozen droplets. Sleet mainly occurs ahead of warm fronts during winter months.

## Degrees of precipitation

Stratiform (large-scale) precipitation occurs as a consequence of the slow ascent of air in synoptic systems, such as along cold fronts, and in advance of warm fronts. This slow air ascent is also seen around Tropical cyclones outside of the eyewall, and in comma head precipitation patterns around mid-latitude cyclones.

Convection rain (showery precipitation) occurs from convective clouds—cumulonimbus or cumulus congestus—and falls as patchy showers with rapidly changing intensity. This type of precipitation is seen most often in the tropics, and is characterized by graupel and hail. In mid-latitudes, convective precipitation generally occurs behind cold fronts, or in association with squall lines and warm fronts with significant available moisture.

Orographic precipitation occurs on the windward side of mountains located near a large-scale source of moist air that flows across the mountain ridge, resulting in adiabatic cooling and condensation. This phenomenon is most often seen on mountainous, oceanic islands, where windward areas are characterized by heavy rainfall, and leeward sides are very dry.

The tropics experience large air masses several hundred miles across with central low pressures and blowing winds. Precipitation in these areas occurs when a warm front moves up an inclined surface of retreating cold air and is cooled, producing rain.

## COMPETENCY 14.0　Understand characteristics of broad-scale weather systems and local weather, the relationship between them, and the methods and instruments used to collect weather data.

## SKILL 14.1　Identify the types and characteristics of air masses, their movements, and the kinds of fronts that form between air masses.

**Parcel of Air (Air Mass):** a large area of air that assumes a characteristic temperature, pressure, and humidity from sitting over a landmass. Horizontally, the parcel has relatively uniform temperature, pressure, and humidity. Vertically, the parcel may have widely differing temperatures, pressures, and humidity.

**Meteorological Terms**

**Isobar**: a line on a meteorological map that connects locations of equal pressure.

**Front:** a narrow zone of transition between air masses of different densities that is usually due to temperature contrasts. Because they are associated with temperature, fronts are usually referred to as either warm or cold.

**Warm Front**: a front whose movement causes the warm air (less dense) to advance, while the cold air (more dense) retreats. A warm front usually triggers a cloud development sequence of cirrus, cirrostratus, altostratus, nimbostratus, and stratus. It may result in an onset of light rain or snowfall immediately ahead of the front, which gives way to steady precipitation (light to moderate) as the cloud sequence forms, until the front passes, a time frame that may exceed 24 hours. The gentle rains associated with a warm front are normally welcomed by farmers. However, if it is cold enough for snow to fall, the snow may significantly accumulate. If the air is unstable, cumulonimbus clouds may develop, and brief, intense thunderstorms may punctuate the otherwise gentle rain or snowfall.

**Cold Front**: a front whose movement causes cold air (more dense) to displace warm air (less dense). The results of cold front situations depend on the stability of the air. If the air is stable, nimbostratus and altostratus clouds may form, and brief showers may immediately precede the front. If the air is unstable, there is greater uplift, cumulonimbus clouds may tower over nimbostratus clouds. Cirrus clouds are blown downstream from the cumulonimbus clouds by high altitude winds. Thunderstorms may occur, accompanied by gusty surface winds and hail, as well as other, more violent weather. If the cold front moves quickly (roughly 28 mph or greater), a squall line of thunderstorms may form either right ahead of the front or up to 180 miles ahead of it.

**Occluded Front**: a front where a cold front has caught up to a warm front and has intermingled, usually by sliding under the warmer air. Cold fronts generally move faster than warm fronts and occasionally overrun slow moving warm fronts. The weather ahead of an occluded front is similar to that of a warm front during its advance, but switches to that of a cold front as the cold front passes through.

**Stationary Front**: a front that shows no overall movement. The weather produced by this front can vary widely and depends on the amount of moisture present and the relative motions of the air pockets along the front. Most of the precipitation falls on the cold side of the front.

### SKILL 14.2    Identify the horizontal and vertical movements of air in high- and low-pressure areas.

**Air Pressure**: the force exerted by the atmosphere. Air pressure is important because it affects the airflow laterally (horizontally). Temperature affects the airflow vertically.

Pressure is a function of temperature and density and is based upon the basic **Gas Law**: Pressure = Constant x Temperature x Density.

If you change one variable, the other variables respond. In the atmosphere both variables (temperature and density) are changing, so the situation is complex.

Air pressure is basically the weight of the overlaying atmosphere. As you rise in altitude, there is less pressure (weight) above you. Temperature increases indicate a greater molecular energy, but although the energy increases, density decreases.

**Measuring Air Pressure**

Mercurial and Aneroid Barometers are used to measure air pressure. Surface pressure is measured using the **mercurial barometer**. An **aneroid barometer** is used to measure pressure at altitudes in the atmosphere. Both types of barometers measure the pressure exerted on the column of mercury, and the measurement is expressed in inches of rise or fall of the mercury. Rising pressure indicates fair weather; falling pressure indicates rain (foul weather.) The average surface pressure is 29.92 inches at sea level.

Meteorologists measure pressure changes in millibar (mb) units, where 1 mb = 0.02953 inches of rise or fall of the mercury. The average air pressure ranges from 950 to 1050 mb. The average air pressure is 1013.25 mb at sea level. Example: The lowest recorded sea level pressure, 870 mb (25.69 in.), was during Typhoon Tip in 1979. The highest recorded sea level pressure was 1083.3 mb (32.01 in.), at Agata, Siberia in 1968.

## High and Low Pressure Areas

Surface pressure is affected by temperature and reflects what is happening in the air. This is very useful for predicting changes in the weather. Temperature plays a large part in the changes in upper atmosphere air pressure.

Pressure is the force exerted by molecules in the air. If you take two columns of air (A & B), with the same temperature and same density, and raise the temperature of B and lower the temperature of A, then you will cause an expansion or contraction of the columns of air.

- Column A contracts because you are lowering the temperature. The molecules become less energetic and are packed closer together, hence, denser.

- Column B expands because you've raised the temperature. The molecules become more active and are packed more loosely, hence, less dense.

- Although the columns now are of physically differing sizes (volume), the pressure is identical in both columns because the density has adjusted for the difference in volume.

- However, the atmosphere is not neatly isolated into columns. Another way to consider this pressure change is to visualize two columns of air not isolated from each other.

- Columns A and B both have the same temperature, density, and pressure throughout. Now, cool the air in column A and heat the air in column B. Overlay these two columns of air with a scale calibrated in millibars. The air in column B now has a higher pressure than the air in column A when both are at the same altitude.

Meteorogically, the air in column A represents an area of **low pressure**, and the air in column B represents an area of **high pressure**. Since nature abhors a vacuum, the air tries to move from the area of high pressure to the area of low pressure. Surface pressure in column A starts to rise because the air is moving. Surface pressure in column B starts to drop because the air is leaving.

When predicting weather, the effect of temperature is more important to changes in upper atmosphere air pressure than to changes in the surface pressure. Temperature has the biggest effect on air pressure in the atmosphere.

Weather maps reflect areas of high and low pressure. In general, the undulating surface is lower to the North and higher to the South. However, the surface has waves in it that form troughs and ridges. These correspond to troughs of low pressure and ridges of high pressure. Surface lows are always offset to the right.

**The Effect of Water on Pressure**

The moisture in the air affects pressure and the effect is more prevalent in the surface pressure. Water vapor in the air makes the air less dense. Example: The atomic weight of an air molecule is 29 amu (Atomic Mass Units). The atomic weight of a water molecule is 18 amu. For every molecule of air displaced by water, the density of the air decreases by 7 amu.

A change in density affects the pressure of the air and all the factors of the Gas Law come into play since temperature and density both affect pressure.

Dry, cold air has the highest pressure, while warm, moist air has the lowest pressure.

**SKILL 14.3**   **Describe the use of weather instruments (e.g., thermometer, psychrometer) for collecting given types of weather data.**

Humidity is measured using a **hygrometer**. There are four basic types.

**Hair hygrometer**: actually uses human hair. Based on the scientific principle that hair lengthens or shortens with a change in humidity. When someone talks about having a bad hair day, they aren't exaggerating. The humidity causes hair to frizz up or go limp.

**Electrical hygrometer:** Based on the principle that electrical conductivity changes with a change in humidity.

**Infrared hygrometer**: Based on the amount of infrared energy absorbed by the air. Changes in the absorption value signal changing humidity.

**Dew Cell**: Based on the principle of measuring the pressure of water vapor.

A **psychrometer** is used to measure relative humidity. This device is also referred to as the wet/dry bulb method, because the psychrometer uses a wet bulb and a dry bulb. Based on the principle of measuring evaporative cooling.

A **sling psychrometer** is the most commonly used device. When whirled around in the air it gives two readings - the temperature of the dry bulb and the temperature of the wet bulb. The difference between the temperatures is indexed on a psychrometer table to find the percentage of relative humidity.

Mercurial and Aneroid Barometers are used to measure air pressure. Surface pressure is measured using the **mercurial barometer**. An **aneroid barometer** is used to measure pressure at altitudes in the atmosphere. Both types of barometers measure the pressure exerted on the column of mercury, and the measurement is expressed in inches of rise or fall of the mercury. Rising pressure indicates fair weather. Falling pressure indicates rain (foul weather). The average surface pressure is 29.92 inches at sea level.

Temperature is a measure of hotness, not heat. Although we use **thermometers** to measure temperature, we can't directly measure heat; it must be calculated.

Temperature is measured on three scales. These are degrees Fahrenheit (F), degrees Celsius (C), and Kelvin units (K). Note: Kelvin (K) is the measurement unit used in the System International (SI).

Celsius Scale: the temperature scale in general scientific use. Water freezes at 0° and boils at 100° on the Celsius scale.

Fahrenheit Scale: the non-SI technical scale still used in some countries. Water freezes at 32° and boils at 212° on the Fahrenheit scale.

Kelvin (K): the SI base unit of temperature, Kelvin is a unit on an absolute temperature scale. Note: There is no degree symbol (°) written after a temperature expressed in Kelvin.

The point at which all motion ceases is called **Absolute Zero**. This occurs at 0 Kelvin, -273 degrees Celsius, and -459.4 degrees Fahrenheit. Note: Absolute Zero does not represent the temperature of freezing, which is 32 degrees Fahrenheit, 0 degrees Celsius, and 273 Kelvin.

**COMPETENCY 15.0  Understand weather maps, how they are prepared, and how they are used.**

**SKILL 15.1  Interpret symbols used on weather maps.**

**Meteorological Map Symbology**

Meteorology utilizes an extensively defined set of alphanumeric notations and iconology to graphically display the collected data elements. Depending on the extent of detail desired in the presented map, the symbology conveys information about cloud types, coverage and base heights, visibility, precipitation, wind speed, air pressure tendency, temperature, and isobaric fronts.

Once you can read a station plot you can begin to perform map analyses. Meteorologists use the station plots to draw lines of constant pressure (isobars), temperature (isotherms), and dewpoint (isodrosotherms) to achieve an understanding of the current state of the atmosphere. This knowledge ultimately leads to better weather forecasts and warnings.

Decoding these plots is easier than it may seem. The values are located in a form similar to a tic-tac-toe pattern.

In the upper left, the temperature is plotted in Fahrenheit. In this example, the temperature is 77°F.

Along the center, the cloud types are indicated. The top symbol is the high-level cloud type followed by the mid-level cloud type. The lowest symbol represents low-level cloud over a number that tells the height of the base of that cloud (in hundreds of feet). In this example, the high level cloud is Cirrus, the mid-level cloud is Altocumulus and the low-level cloud is a cumulonimbus with a base height of 2000 feet.

At the upper right is the atmospheric pressure reduced to mean sea level in millibars (mb) to the nearest tenth with the leading 9 or 10 omitted. In this case the pressure would be 999.8 mb. If the pressure was plotted as 024 it would be 1002.4 mb. When trying to determine whether to add a 9 or 10 use the number that will give you a value closest to 1000 mb.

On the second row, the far left-hand number is the visibility in miles. In this example, the visibility is 5 miles.

Next to the visibility is the present weather symbol. There are 95 symbols that represent the weather that is either presently occurring or has ended within the previous hour. In this example, a light rain shower was occurring at the time of the observation.

The circle symbol in the center represents the amount of total cloud cover reported in eighths. This cloud cover includes all low, middle, and high level clouds. In this example, 7/8th of the sky is covered with clouds.

This number and symbol tell how much the pressure has changed (in tenths of millibars) in the past three hours and the trend in the change of the pressure during that same period. In this example, the pressure was steady and then fell becoming 0.3 millibars LOWER than it was three hours ago.

These lines indicate wind direction and speed rounded to the nearest 5 knots. The longest line, extending from the sky cover plot, points in the direction that the wind is blowing **from**. Thus, in this case, the wind is blowing **from** the southwest. The shorter lines, called barbs, indicate the wind speed in knots (kt). The speed of the wind is calculated by the barbs. Each long barb represents 10 kt with short barbs representing 5 kt. In this example, the station plot contains two long barbs so the wind speed is 20 kt.

The 71 at the lower left is the dewpoint temperature. The dewpoint temperature is the temperature the air would have to cool to become saturated, or in other words reach a relative humidity of 100%.

The lower right area is reserved for the past weather, which is the most significant weather that has occurred within the past six hours excluding the most recent hour.

**Analyze a Map on Your Own**

The following are a few sources of current weather maps. Sometimes a site may be down or experiencing data losses. In such a case, try another site listed. This is not meant to be an exhaustive list. These are provided for your convenience.

- NCAR, pick your regional plot: http://www.rap.ucar.edu/weather/surface/
- UNISYS: http://weather.unisys.com/surface/sfc_map.html
- College of DuPage: http://weather.cod.edu/analysis/analysis.sfcplots.html
- NOAA   http://www.nws.noaa.gov/
- Ohio State University: http://asp1.sbs.ohio-state.edu/ (Click on "Current Weather" and choose your map)

## Surface Station Model

| Temp (F)<br>Weather<br>Dewpoint (F) |  | Pressure (mb)<br>Sky Cover<br>Wind (kts) | **Data at Surface Station**<br>Temp 45 °F, dewpoint 29 °F, overcast, wind **from** SE at 15 knots, weather light rain, pressure 1004.5 mb |
|---|---|---|---|

## Upper Air Station Model

| Temp (C)<br><br>Dewpoint (C) |  | Height (m)<br><br>Wind (kts) | **Data at Pressure Level - 850 mb**<br>Temp -5 °C, dewpoint -12 °C, wind **from** S at 75 knots, height of level 1564 m |
|---|---|---|---|

## Forecast Station Model

| Temp (F)<br>Weather<br>Dewpoint (F) | 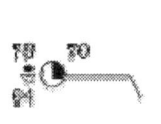 | PoP (%)<br>Sky Cover<br>Wind (kts) | **Forecast at Valid Time**<br>Temp 78 °F, dewpoint 64 °F, scattered clouds, wind **from** E at 10 knots, probability of precipitation 70% with rain showers |
|---|---|---|---|

**Map Symbols**

| Sky Cover | Wind |
|---|---|
| ○ clear | ○ Calm |
| 1/8 | — 1-2 knots (1-2 mph) |
| ◔ scattered | 3-7 knots (3-8 mph) |
| 3/8 | 8-12 knots (9-14 mph) |
| 4/8 | 13-17 knots (15-20 mph) |
| 5/8 | 18-22 knots (21-25 mph) |
| ● broken | 23-27 knots (26-31 mph) |
| 7/8 | 48-52 knots (55-60 mph) |
| ● overcast | 73-77 knots (84-89 mph) |
| ⊗ obscured | 103-107 knots (119-123 mph) |
| missing | Shaft in direction wind is coming **from** |

**SKILL 15.2    Describe the methods used to generate weather maps.**

Weather maps provide graphical or pictorial images of weather to meteorologists and everyday people. There are several different types of weather maps including satellite, radar, front, precipitation, wind speed and direction, temperature, cloud, and pressure trend.

## Satellite

These weather maps are generated using data collected from satellites, such as that of NOAA (National Oceanic and Atmospheric Administration or National Weather Service). These maps display images taken in the infrared band of light and show relative warmth of objects. Satellite weather is mostly used to observe cloud formation. Colder objects are displayed as brighter patches, where warmer objects are displayed as darker patches. Lower layers of clouds are generally warmer and appear gray on weather maps. Colder, higher clouds are highlighted using a color system that portrays their relative temperatures and altitudes. This color system is defined by the weather map's legend.

## Radar

Weather radar is used to locate precipitation, calculate its motion, estimate its type, and forecast its future position and intensity. A pulse of energy is sent through a cloud and the amount of echo returned will give the intensity of its precipitation. The echo is actually a reflection of the energy and a computer will generate a color code to indicate the amount of precipitation. Doppler radar, used by meteorologists, is capable of detecting the motion of rain droplets in addition to intensity of the precipitation. Radar maps are generated using data acquired by Doppler sources. These maps generally portray the location of areas of precipitation, as well as its intensity, using the color legend displayed below.

## Radar Intensity Colors on a Weather Map

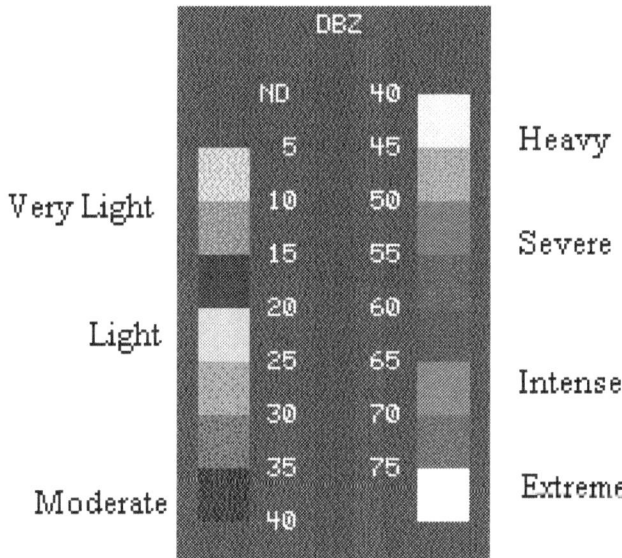

Courtesy of (http://weather.about.com/od/forecastingtechniques/ss/mapsymbols_7.htm).

To generate the following weather maps, meteorologists collect weather data daily at multiple locations using high altitude weather balloons, weather equipment and gauges, satellites, and computers. This information is then converted into maps with weather symbols.

**Fronts**

Meteorologists use an extensive system of symbols to represent a large variety of weather events. Weather fronts are the boundaries between air masses with different characteristics such as temperature, humidity, and air pressure. Movement of a front depends largely on the conditions inside its air mass. Fronts are portrayed on weather maps as follows:

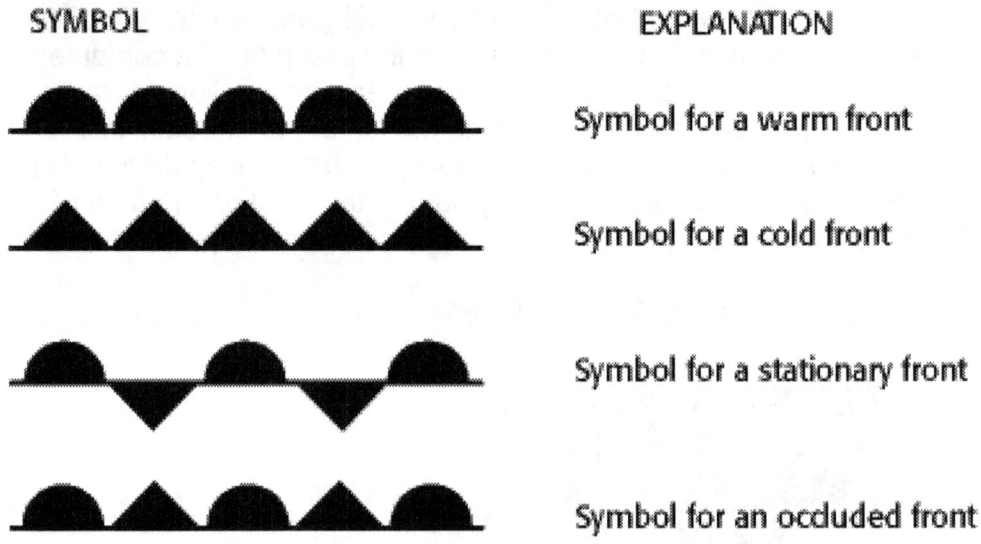

Courtesy of (http://nsidc.org/arcticmet/factors/pressure.html).

## Precipitation

The following diagram provides the weather map symbols for rain, snow, sleet, hail, etc.

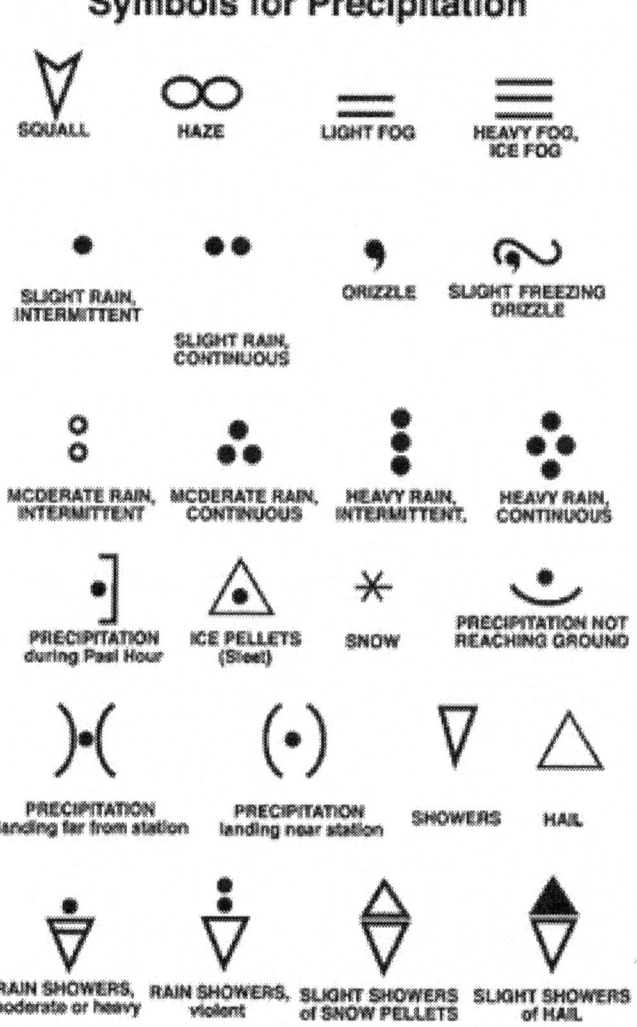

Courtesy of
(http://weather.about.com/od/forecastingtechniques/ss/mapsymbols_6.htm).

**Wind Speed and Direction**

Wind barbs are used to show wind intensity with a series of small pennants. Wind direction is indicated by the compass direction in which the wind diagram points. Wind speed is measured in knots, where 1 knot = 1.1507788 miles per hour.

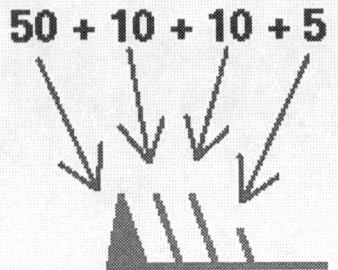

Wind blowing from the west at 75 knots

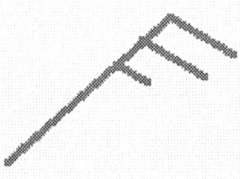

Wind blowing from the northeast at 25 knots

Wind blowing from the south at 5 knots

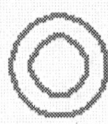

Calm winds

Courtesy of (http://www.hpc.ncep.noaa.gov/html/stationplot_buoy.shtml).

**Temperature**

Weather maps generally represent temperature with actual numbers instead of symbols. This method can be seen on the following map.

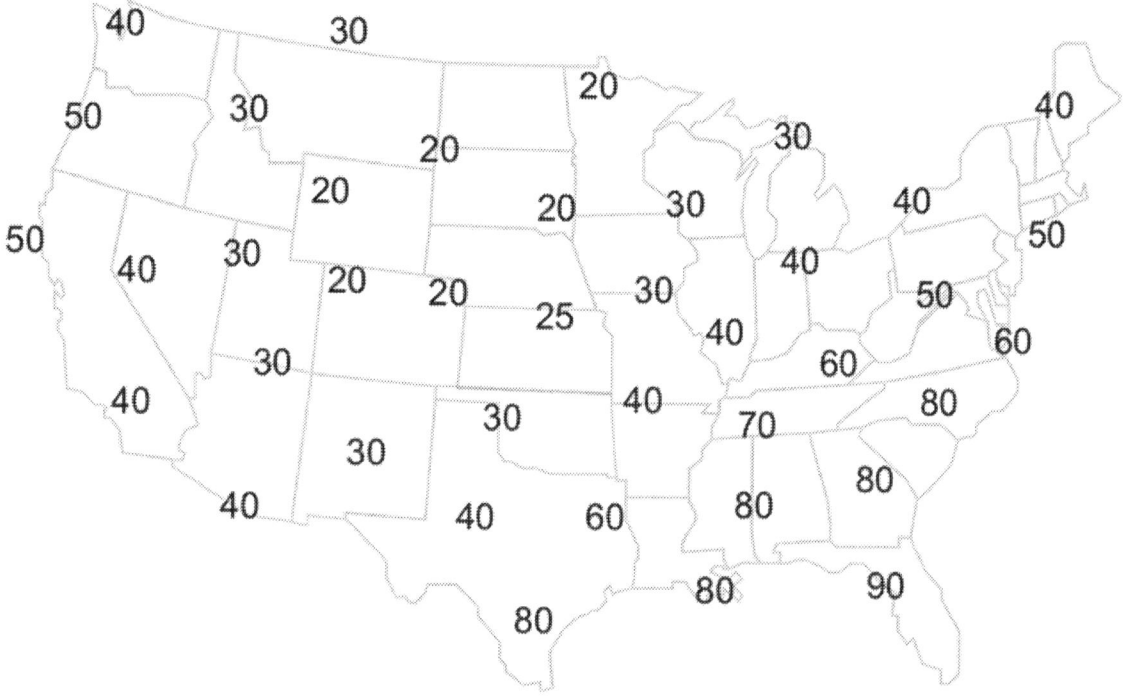

Courtesy of (http://www.srh.weather.gov/srh/jetstream/synoptic/temp.htm).

## Cloud Cover

In addition to radar images of clouds, weather maps are also generated using the following cloud cover symbols. The amount that the circle is filled at the center of the station plot reflects the approximate amount of the sky that is covered by clouds.

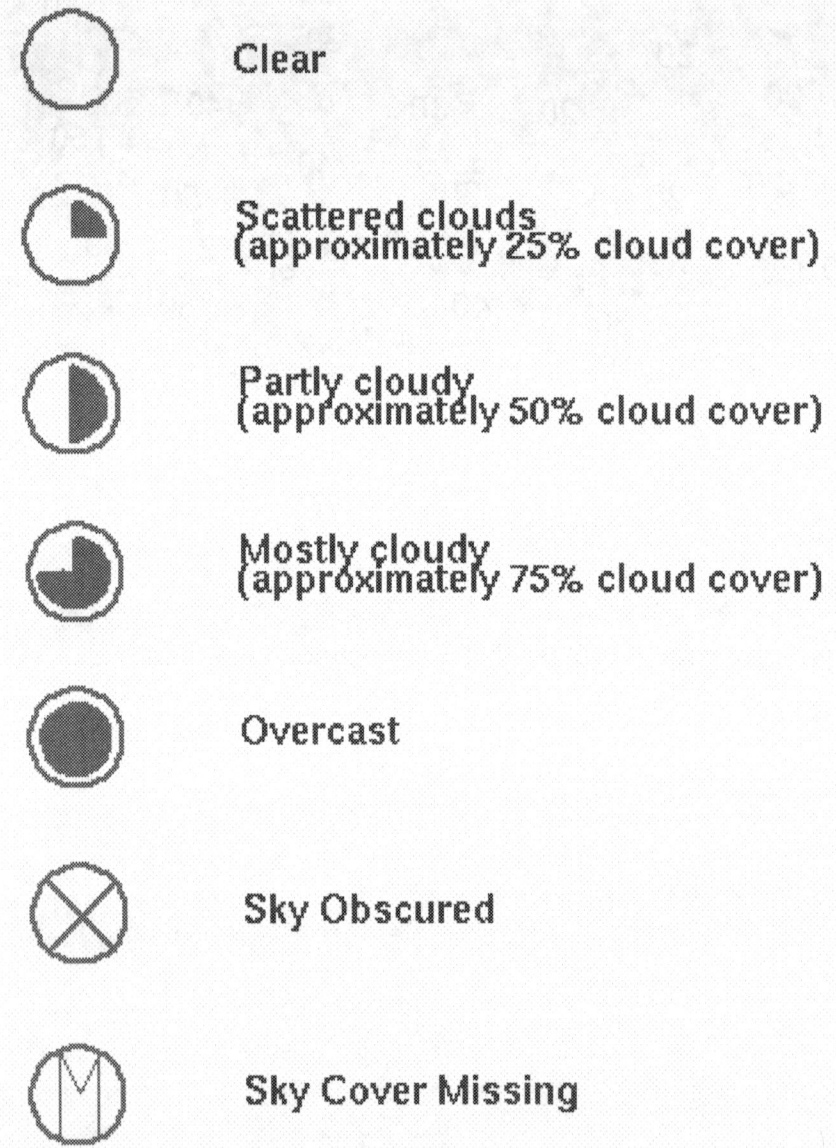

Courtesy of (http://www.hpc.ncep.noaa.gov/html/stationplot_buoy.shtml).

## Pressure

A particular pressure trend is represented on a weather map with a number and a symbol. The number provides the 3-hour change in tenths of millibars, while the symbol provides a graphic illustration of how this change occurred.

Courtesy of (http://www.hpc.ncep.noaa.gov/html/stationplot).

**SKILL 15.3**     Infer recent weather in a given location based on one or more weather maps.

Using symbols, the following map portrays the location of weather stations, as well as the temperature, dew point, and pressure found near a particular weather station. Cloud coverage, precipitation type, wind speed, wind direction, and weather fronts are also depicted on this map. Weather stations are represented by circular symbols.

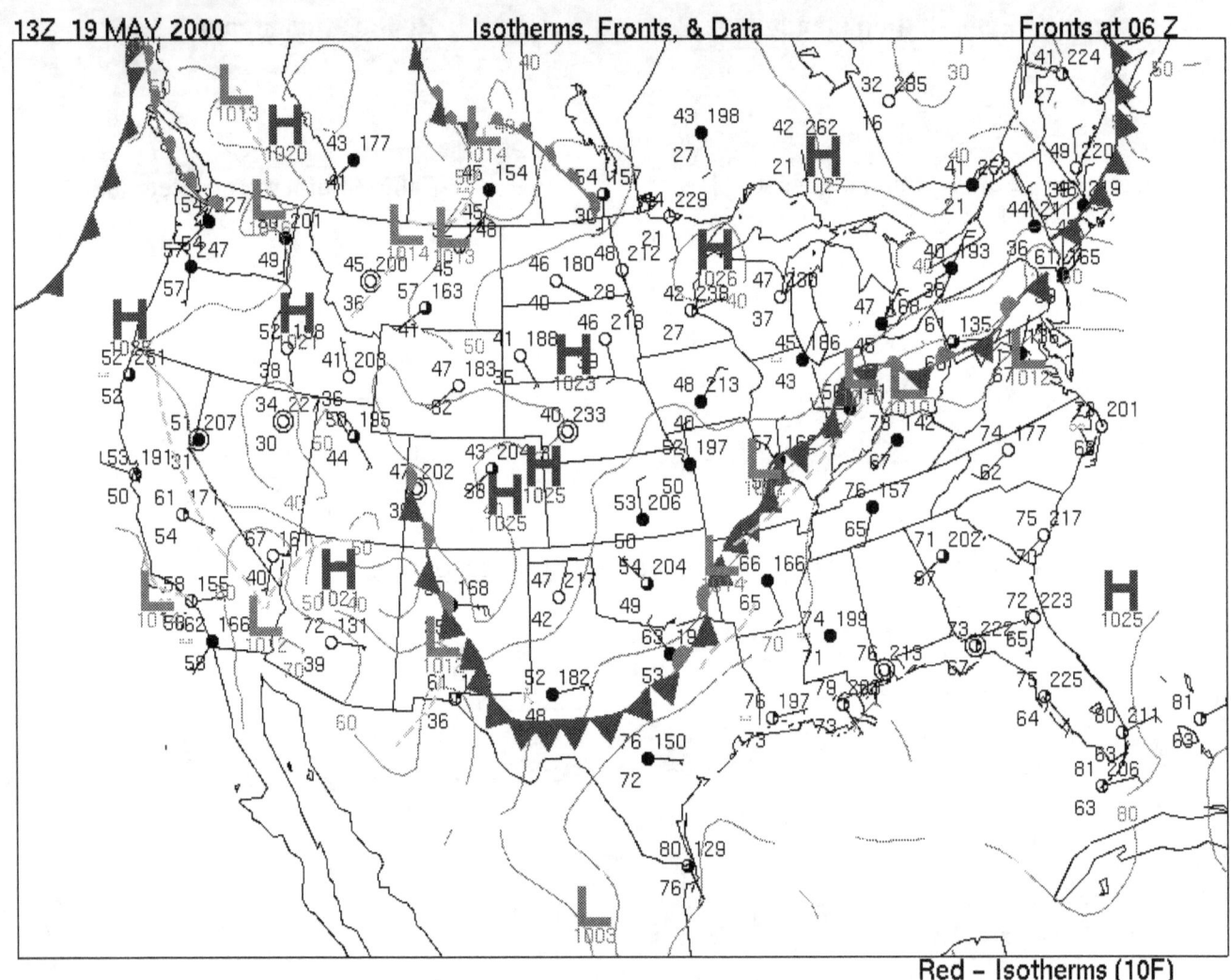

http://www.dnr.sc.gov/climate/sercc/education/saer/aer_summer_00.html

## Temperature, Dew Point, and Pressure

The value for temperature is located to the upper left side of the circular weather station symbol, dew point to the lower left, and pressure to the upper right of the station symbol. According to this map, the weather station in Western Texas is recording a temperature of 52 degrees F and a dew point of 48 degrees F. The number to the upper right of the weather station symbol represents the last three digits of the pressure recorded in this area. If this recorded value is greater than 500, the initial 9 is missing. If the recorded value is less than 500, the initial 10 is missing. The pressure is determined by placing the 9 or 10 in front of the reported value and dividing by 10. Therefore, the pressure number of 182 for Western Texas indicates a pressure of 1018.2.

## Front

The blue, flag-like symbols in this map indicate a cold front moving southward away from the weather station.

## Cloud Coverage

Cloud coverage in Western Texas is portrayed by the degree to which the weather station circle is filled in. According to this map, skies over this weather station are overcast.

## Wind Speed and Direction

Wind speed is shown on a weather map by various symbols. Calm conditions are depicted by two open circles, one inside the other. Winds 1-2 mph are depicted by a straight line, and wind speeds above 1-2 are depicted by one or more barbs and/or triangles attached to a straight line. The wind speed chart is as follows:

| Winds | mph |
|---|---|
| ◎ | Calm |
| —— | 1-2 |
| —⼃ | 3-8 |
| —＼ | 9-14 |
| —＼₁ | 15-20 |
| —＼＼ | 21-25 |
| —＼＼₁ | 26-31 |
| —＼＼＼ | 32-37 |
| —＼＼＼₁ | 38-43 |
| —＼＼＼＼ | 44-49 |
| —＼＼＼＼₁ | 50-54 |

| Winds | mph |
|---|---|
| ▲— | 55-60 |
| ▲⼃— | 61-66 |
| ▲＼— | 67-71 |
| ▲＼₁— | 72-77 |
| ▲＼＼— | 78-83 |
| ▲＼＼₁— | 84-89 |
| ▲▲⼃— | 119-123 |

The direction from which the long, flat wind line enters the weather station symbol is the direction from which the wind blows. Therefore, this station is reporting winds of 15-20 mph, blowing from the East–Northeast.

## Precipitation

Type of precipitation is indicated by the following legend:

| Current Weather | Sky Coverage |
|---|---|
| • Intermittent Rain | ◯ No Clouds |
| •• Continuous Rain | Less than one-tenth or one tenth |
| ✶✶ Continuous Snow | Two-tenths or three-tenths |
| ,, Continuous Drizzle | Four-tenths |
| ≡ Fog | Five-tenths |
| ↗ Thunderstorm | Six-tenths |
| ↗̇ Thunderstorm with Intermittent Rain | Seven-tenths or eight-tenths |
| | Nine-tenths or overcast with openings |
| | Completely overcast |
| | ⊗ Sky obscured |

http://www.dnr.sc.gov/climate/sercc/education/saer/aer_summer_00.html

Therefore, in this map, Western Texas is experiencing thunderstorms with intermittent rain.

# TEACHER CERTIFICATION STUDY GUIDE

**SKILL 15.4**     **Predict future weather in a given location based on one or more weather maps.**

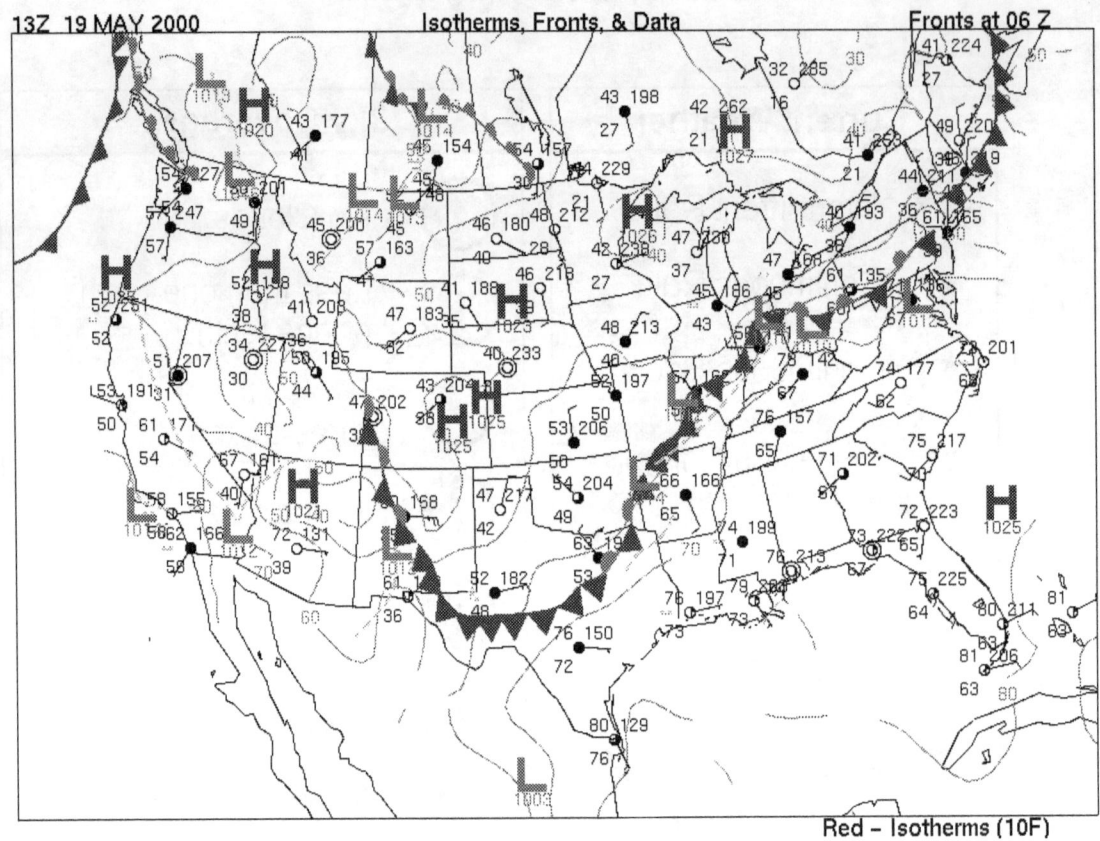

http://www.dnr.sc.gov/climate/sercc/education/saer/aer_summer_00.ht

This map can be used to predict future weather conditions at the weather station in central Arkansas represented by the circle symbol.

This map indicates that a cold front is moving toward the weather station from West to East, as do most weather conditions in the United States. Cold weather fronts occur when a cold air mass of high density pushes under a warm air mass, causing moisture in the warm air to quickly rise, cool, and condense. Heavy precipitation in the near future can be predicted for this area.

This weather map indicates that a low-pressure system is moving toward the weather station with the approaching cold front. Low-pressure systems allow air to rise and moisture to condense, and are associated with precipitation. The cloud coverage symbol of a completely dark circle indicates that skies are overcast. This state could be due to the presence of nimbostratus clouds, which form dark sheets that block sunlight and are often responsible for lengthy periods of precipitation. Therefore, all conditions for Arkansas portrayed on this weather map indicate imminent precipitation.

**EARTH SCIENCE**

# TEACHER CERTIFICATION STUDY GUIDE

**COMPETENCY 16.0**     **Understand the principles and technology of weather forecasting and the effects of weather and weather forecasting on humans.**

**SKILL 16.1**     **Describe the use of weather models in forecasting.**

Weather forecasting is the prediction of the future state of the atmosphere through the use of science and technology. After the processes of data collection and assimilation, numerical weather prediction (NWP) models are often used to produce accurate future weather scenarios through mathematical model computer programming. NWP focuses on two distinct problems, diagnosing the current state of the atmosphere and numerically modeling how the atmosphere will change over time. This type of weather model is capable of manipulating large data sets and performing the complex calculations needed to analyze weather observation data. Numerical weather prediction models create computer simulations of the atmosphere using an understanding of physics and fluid dynamics. A numerical weather prediction model is run once using a given set of initial conditions to produce a single forecast. To correct for forecast errors due to the chaotic nature of the atmosphere, an ensemble suite of forecasts may be run with each forecast having a slightly different set of initial conditions. This type of modeling generates a range of different weather scenarios, and may be used to calculate the probability for a particular forecast sequence.

Earth's climate is largely influenced by the water budget and a better understanding of water flux greatly increases weather forecasting capabilities and accuracy. For this reason, the Global Hydrology and Climate Center is currently working on a project that will improve scientists' ability to diagnose global distributions of water vapor, cloud condensation, and precipitation by developing a water budget diagnostic model measured around the globe as a function of time. This model uses wind, temperature, and initial moisture fields from a global four-dimensional gridded data set as input to produce equations that calculate future water vapor, cloud condensation, and precipitation values. Using these values, scientists are able to predict the movement of water in the atmosphere, and thus the formation of precipitation and other types of weather. The Global Hydrology and Climate Center's project demonstrates how modeling greatly improves weather forecasting.

**SKILL 16.2**     **Identify the role of computers and satellite photographs**

**Weather Stations**

The National Weather Service runs the 500 active weather stations used for national forecasting. However, there are thousands of local weather stations that contribute observations. They collect temperature, precipitation, and humidity data and provide remarks about local conditions.

Data is collected 4 times daily, every 6 hours, around the clock, and sent to the National Weather Bureau. They in turn provide it to the World Meteorological Association, who use it for making long-term predictions, not immediate forecasting.

Meteorological data is collected on the basis of GMT time (Zulu Time). **GMT** refers to Greenwich Mean Time. Greenwich, England is located on the prime meridian, or 0 degrees. Time is measured as either plus or minus the current GMT. Each line of longitude is equal to 4 minutes of time and 15 degrees of longitude is equal to 1 hour of time. You subtract hours as you go west, and add hours as you go east. Example: 8 AM Zulu is equal to 6 AM at 30° west longitude.

**Ground Based Radar**: "Doppler Effect" radar is the primary type used because it is very accurate, sensing motion as well as range and direction. This allows for pinpoint accuracy over an area. Example: This is the type of radar you see on television. The **NEXRAD** Radar. (Next Generation Doppler Radar) emits beams of energy that are reflected by the water droplets in the atmosphere. This type of radar is very useful for tracking and predicting rain and less useful for snow or sleet.

**Satellites**: The National Weather Service heavily depends on its network of weather satellites to provide wide-area coverage of the Earth. These satellites primarily provide infrared, water vapor, and photographic data and are used to track the formation, development, and motion of major meteorological events such as hurricanes and tropical storms.

In terms of orbit, there are two types of satellites: Geostationary and Polar Orbiting. **Geostationary satellites** move with the Earth's rotation. Since they always look at the same point, this allows for a view that shows changes over periods of time. **Polar Orbiting satellites** follow an orbit from pole to pole. The Earth rotates underneath the satellite and gives a view of different areas. In effect, it produces slices of the Earth.

The **GOES Satellite** is a key source of weather information. GOES is a geostationary satellite that scans the Atlantic Ocean and the U.S. East Coast. There are 4 detection bands on the GOES satellite.

**1. Visible Spectrum**: Photographic surveys. Pictures are actually light reflecting off of the tops of clouds. This capability is limited to daytime use because there is no reflected light during the nighttime hours.

**2. Near IR**: Looks at heat. Emphasis is on detecting naturally emitted and reflected infrared radiation.

**3. Enhanced IR**: Adds colors to the tops of the clouds. Uses an enhanced IR (Infrared) receiver/detector to look at reflected or emitted heat. Hotter areas show up as red/yellow/orange, while colder areas appear as blue/green/violet.

**4. Water Vapor**: Uses radar to detect the water vapor in the atmosphere. The swirls produced by this type of satellite show the air rising. It looks for uplifts that provide an indicator of weather change. The higher the cloud, the colder it is. The rising clouds can indicate thunderstorms, and high, well-developed clouds can signal the advance of a storm formation.

## Using the Collected Data

The collected data is sent to the National Meteorological Center in Maryland where it is plotted and then sent on to other weather centers throughout the world. The collected data is used to develop mathematical models that are used to predict the weather. These models have been used since the late 1950's and place differing emphasis on the various aspects of the equational factors involved. A typical model will consist of 6-8 different equational factors.

In turn, the data elements are disseminated to the National Weather Service, which uses them to calculate model results based on 12, 24, 36, 48, and 72-hour predictions. However, the further out you extend the predictions, the less accurate the prediction is likely to be. This is because of the wide variation in the data elements that can occur between the time of prediction and the realization of the predicted time.

The data model is also disseminated to local meteorological offices and radio and television stations for local interpretation. The resulting local forecast often includes a great deal of **Kentucky Windage**: experience based guesses by local meteorologists, and can vary considerably from the forecasts of the National Weather Service. The accuracy of local forecasting is dependent on the experience and expertise of the local meteorologist. Forecasting is only as good as the data collected and local intuition (experience) modifications. Four main types of forecast are developed:

**1. Persistence**: This doesn't really change much and is very inaccurate.

**2. Steady State (Trend):** This is a guess as to what the system will do based on the assumption that there will be no change in the data parameters.

**3. Analog**: This is the best type of forecast basis. It takes the data numbers and plugs them into equations, the equations into models, and uses the models for forecasting. This is also called Numerical Forecasting.

**4. Climatological**: This type of forecast is often vague and is based on long-standing trends. Example: The weather predictions in the "Farmer's Almanac."

## Forecasting Language

### Cloud Forecast

| Forecasted | % Cloud Coverage |
|---|---|
| Fair | < 40% |
| Partly Cloudy/Partly Sunny | 30 to 70% |
| Cloudy | 90% |

### Precipitation Forecast

| Forecasted | %Probability |
|---|---|
| Slight Chance | 10 to 20% |
| Chance | 30 to 50% |
| Occasional | > 50%, but for only ½ of the time period covered |

**SKILL 16.3**      **Describe types of hazardous weather.**

**Tornadoes**

**Tornado**: an area of extreme low pressure, with rapidly rotating winds beneath a cumulonimbus cloud. Tornadoes are normally spawned from a Super Cell Thunderstorm. They occur when very cold air and very warm air meet, usually in the Spring.

Tornadoes represent the lowest pressure points on the Earth and move across the landscape at an average speed of 30 mph. The average size of a tornado is 100 yards, but they can be as large as a mile wide. A tornado's wind speed ranges from 65 to 300 mph and has an average duration of 10 to 15 minutes, but has been known to last up to 3 hours. Tornadoes usually occur in the late afternoon (3 to 7 p.m.), in conjunction with the rear of a thunderstorm. Most tornadoes spin counter-clockwise in the northern hemisphere and spin clockwise in the southern hemisphere. They are not dependent on the Coriolis Effect because of the way they are formed.

Worldwide, the U.S. has the most tornadoes and most of these occur in the Spring. Texas has the most tornadoes, but Florida has the largest number per square mile. Tornadoes are without a doubt the most violent of all storms. Roughly 120 people each year are killed in the United States by tornadoes.

## Formation of a Tornado

Technically, tornadoes are actually classified as mesocyclones that form within super cell thunderstorms. Temperature differentials within the thunderstorm cause an up and down draft effect that induces a horizontal spinning motion. An increase in the up and down motion tightens the spin and a strong draft may eventually tilt the **mesocyclone** into a vertical position. When this occurs, the bottom of the cloud wall drops down as a funnel cloud. If the funnel cloud touches the ground, it becomes a tornado. Approximately 50% of mesocyclones produce tornadoes.

Tornadoes are very rare in **Pop-Up** and **Squall Line Thunderstorms**. These are both isolated storm fronts of very short duration and speeds. Although there is an induced spin similar to a mesocyclone, it is on a much smaller scale. Tornadoes sometimes form **families**: multiple funnels within the same super cell thunderstorm. If all of these funnels touch the ground, a very rare tornado formation known as a **Rake** can, as the name implies, affect a much wider area in its movement path.

## Detection and Tracking

Doppler Radar is the primary device used in tracking tornadoes. The radar measures the funnel's speed and movement by using false color techniques. Movement is graphically displayed with red/yellow patterns indicating a movement away from the radar, and blue/green patterns indicating movement toward the radar.

Mobile tornado tracking devices include TOTO (Totable Tornado Observatory) which was developed by researchers at the University of Oklahoma at Norman, Turtles, expendable ground sensors that measure temperature and pressure, and portable Doppler radars.

Authorities may issue a **Tornado Watch** if meteorological conditions could/probably will cause the formation of mesocyclones. A **Tornado Warning** is issued when a funnel cloud is spotted. Tornadoes are also noted for producing unusual effects (besides total destruction) in their path.

Observers have routinely reported such phenomena as straws driven into tree trunks, people lifted into the air and then gently returned to the ground miles away, and homes left intact while homes on either side were devastated. For example, while assisting in the cleanup efforts after a tornado struck Topeka, Kansas, in 1967, the author of this book viewed a home that, although it had completely lost its roof and exterior walls, had its interior furnishings untouched by the tornado. Despite the violence visited on the home by the tornado, the furniture and the evening meal on the table had remained in their original positions.

**Waterspouts** and **Dust Devils** are on a smaller scale and are far less powerful versions of mesocyclones. Waterspouts occur only over water, and thermal currents not associated with super cell thunderstorms cause dust devils.

### The Fujita Scale

The **Fujita Scale** is used to measure the intensity and damage associated with tornadoes. Numbered from 0 to 6, ascending numbers on the Fujita Scale indicate higher wind speeds and damage.

| Intensity | Wind (mph) | Typical Damage |
|---|---|---|
| *Weak* | | |
| F0 | <72 | broken branches, shallow trees uprooted, damages signs and chimneys |
| F1 | 72-112 | damage to roofs, moving auto swept off road, mobile homes overturned |
| *Strong* | | |
| F2 | 113-157 | roofs torn off homes, mobile homes completely destroyed, large trees uprooted |
| F3 | 158-206 | trains overturned, roofs and walls torn off of well constructed houses |
| *Violent* | | |
| F4 | 207-260 | frame houses completely destroyed, cars picked up and blown windward |
| F5 | 261-318 | steel reinforced concrete structures badly damaged |
| F6 | >319 | might occur in a small part of an F4 or F5 tornado. Damage would be indistinguishable from that of the main body of the tornado. |

### Hurricanes

**Hurricanes** are produced by temperature and pressure differentials between the tropical seas and the atmosphere.

Powered by heat from the sea, they are steered by the easterly trade winds and the temperate Westerlies, as well as their own incredible energy. Hurricane development starts in June in the Atlantic, Caribbean, and Gulf of Mexico, and lasts until the end of the hurricane season in late November.

Hurricanes are called by different names depending on their location. In the Indian Ocean they are called **Cyclones**. In the Atlantic, and east of the international dateline in the Pacific, they are called Hurricanes. In the western Pacific they are called **Typhoons**. Regardless of the name, a hurricane can be up to 500 miles across, last for over two weeks from inception to death, and can produce devastation on an immense scale.

**Formation of a Hurricane**

Hurricanes start as an upper atmospheric disturbance at the 500mb altitude. This is a trough of low pressure called a **Tropical Wave**.

**Recipe for a hurricane**

Warm water of at least 200 feet in depth and 79° F in temperature.
Converging surface winds.
Diverging upper atmosphere winds, diverging at a faster rate than the surface winds are converging.
High humidity and unstable air (warm air rising).
Moderate strength winds aloft, and they must blow from the same direction to add momentum to the self-propagating effect.

**Warm water feeds hurricanes**. The greater the mass of water available, the greater the spin induced by the winds going in the same direction. The storm essentially rotates counter- clockwise around itself, pulling in more wind in a self-perpetuating effect.

Divergence causes the storm to grow as the condensation of the cloud mass releases heat to further feed the storm. Bands of clear air form between the spiral bands (thunderstorms) of the hurricane and a clear, low-pressure **eye** forms at the center of the storm.

The tightness of the eye indicates the relative strength of the storm. The more pronounced and tight the eye is, the stronger the storm. The highest winds are located next to the eye wall. Although as a general rule, hurricanes move east to west, a hurricane's path is determined by its interaction with areas of high and low pressure cells. These cells steer the storm with lows repelling it, and highs attracting it.

Hurricanes die when they move over land or cold water. The hurricane's energy dissipates slowly when robbed of its water or heat source. Stable air breaks the self-propagating effect.

## Formation Sequence

**Tropical Wave**: a trough of low pressure in the trade wind easterlies, very common throughout the tropics during the Summer and Fall.

**Tropical Disturbance**: a moving area of thunderstorms in the tropics that maintains an identity for a period of 24 hours or more.

**Tropical Depression**: the first sign of a developing storm, with sustained surface winds of 38 mph or less. Tropical Depressions show some rotation and are assigned a number by the National Hurricane Center in Miami, and the storm's progression and development are tracked.

**Tropical Storm**: a tropical cyclone in which the maximum sustained surface winds have reached 39 mph or stronger. The storm has well defined rotation, is assigned a name, and is continuously watched for further development.

**Hurricane**: a tropical cyclone in which the maximum sustained surface winds reach 74 mph or greater. The Northeast (right hand) quadrant of the storm is the strongest because of addition of wind to forward momentum.

## The Saffir-Simpson Scale

The **Saffir-Simpson Scale** is used to classify hurricanes into five categories, with increasing numbers corresponding to lower central pressures, greater wind speeds, and large storm surges.

| Category | Pressure (mb) | Wind Speed (mph) | Storm Surge(m) | Damage |
|---|---|---|---|---|
| 1 | ≥980 | 74-95 | 1-2 | minimal |
| 2 | 965-979 | 96-110 | 2-3 | moderate |
| 3 | 945-964 | 111-130 | 3-4 | extensive |
| 4 | 920-944 | 131-155 | 4-6 | extreme |
| 5 | <920 | >155 | >6 | catastrophic |

## Hurricane Damage

The destruction and damage caused by a hurricane or tropical storm can be severe. **Storm surge** causes most of the damage as the winds push along a wall of rising water in their path, and this rising effect is amplified on low sloping shorelines such as those found on the Gulf Coast. The intense winds can also cause damage. Some notable storms are listed below.

Galveston, Texas, 1900: A 20 feet storm surge killed over 6,000 people and literally destroyed the majority of this Gulf Coast city.

Florida Keys, 1935: A category 5 hurricane, the biggest of the 20th century, was recorded as having 150-200 mph winds and a barometric pressure of 892 mb.

Hurricane Camille, Florida and South Carolina, 1969: This category 5 hurricane with a barometric pressure of 990 mb, scored a direct hit on the Gulf Coast, killing 300 people and causing severe flooding as far north as Virginia.

Hurricane Andrew, Gulf Coast, 1992: Classified as a category 4/5 hurricane, this was the fastest moving storm ever recorded. Moving at 20 mph, the hurricane had maximum sustained winds of 150 mph and gusts up to 175 mph, and re-intensified just prior to making landfall. Andrew was a notable exception to the damage rule in that it was the winds, not the storm surge (16.9 feet), which caused the majority of the 26.5 billion dollars of damage. Over 80,000 houses were destroyed and an additional 50,000 houses suffered greater than 50% damage. 70,000 acres of mangrove swamp was uprooted as this storm cut a 25 mile wide path of destruction in Florida and Louisiana.

**Lightning**

**Lightning**: a brilliant flash of light produced by an electrical discharge of about 100 million volts. Lightning flashes when the attraction between positive and negative charges (ions) becomes strong enough to overcome the air's normally high resistance to electrical flow.

Normally, the surface of the Earth is negatively charged and the upper troposphere is positively charged. However, this distribution changes when a cumulonimbus cloud develops. Charges separate within the cloud so that the upper portion and a small region near the base become positively charged. Likewise, the cloud induces a positive charge on the ground directly beneath it.

As a thunderstorm matures, electrical resistance of the air breaks down and lightning can flow either between oppositely charged areas of the cloud, or between the cloud and the ground. Lightning flashes follow a predictable sequence of events.

Electrons (negatively charged) begin a zigzagging downward path in a forked pattern called a **Step Leader**. When the stepped leader is within 100 meters (328 feet) of the ground it draws a **Return Stroke**, a streamer of positive charge ions sent upward, normally through a tall object such as a tree or flagpole.

When the leader and return stroke meet, a powerful electrical current begins to flow as an intense wave of positively charged ions travel upward at 31,000 miles per second. The entire lightning sequence takes place in less than two tenths of a second.

Following the initial electrical discharge, **Dart Leaders**: additional surges of electrons, flow along the same path and are met by return strokes from the ground. A typical lightning discharge consists of two to four dart leaders plus their return strokes.

Lightning heats the air along the conducting path to temperatures that exceed 45,000° F (25,000° C), and can cause serve burns or death to people near the strike area. This heating also expands the air violently, initiating a sound wave we call **thunder**.

**SKILL 16.4   Identify types and functions of weather precautions.**

Some of the most common natural events are hurricanes, tornadoes, volcanoes, floods, snow and ice, extreme heat, etc. The US government studies these natural hazards and helps people through agencies like FEMA (Federal Emergency Management Agency). FEMA coordinates all of the relief activities so that people can receive assistance as soon as possible.

**Hurricanes**

The term hurricane is derived from Huracan, a god of evil recognized by the Tainos, an ancient aborigines Central American tribe. Hurricanes form over tropical waters (between 8 to 20 degrees Celsius or 80 degrees Fahrenheit or greater). The hurricane season in the northern hemisphere runs from June to November. There are 5 types of hurricanes based on a classification system, which was designed in 1970 by Herbert Saffir, an engineer, and Robert Simpson, the then director of the National Hurricane Center and is called the Saffir-Simpson scale. On this scale 1 is the weakest and 5 is the strongest, referring to the damage sustained. Categories 3,4, and 5 are considered as major (intense) hurricanes capable of inflicting great damage and loss of life. At the center of the hurricane is an area mostly free of clouds, sinking air, and light winds, called the eye. The National Hurricane Center issues warnings about hurricanes and people need to follow them and take necessary precautions including stocking food, filling gas in vehicles, and leaving homes when they are advised to.

**Tornadoes**

A tornado is a rotating column of air extending from a thunderstorm to the ground and is capable of tremendous destruction with wind speeds of 250 mph or more. The path of the tornado can be 1 mile wide and 50 miles long. Tornadoes are the most violent storms and about 1,000 occur each year. There are nearly 80 deaths and 1500 injuries every year due to the destructive effect of tornadoes. They come in all shapes and sizes and they occur mostly in the southern states. The peak season for tornadoes is March to May. Tornadoes are assessed one of five categories depending on the wind speed. The best precaution is to stay indoors when a tornado warning is given and not venture outdoors to observe.

## Volcanoes

Volcanoes are mountains built by accumulation of their own eruptive products - lava, crusted over ash flows, tephra, air borne ash, and dust. Volcanoes are different from other mountains because they are not formed by folding or crumpling or by uplift and erosion. A volcano is most commonly a conical hill or mountain built around a vent that connects with reservoirs of molten rock below the surface of the Earth. The term volcano also refers to the opening of the vent through which the molten rock and associated gases are expelled. Buoyancy and gas pressure drive the molten rock, which is lighter than the surrounding solid rock, forcing it's way upward, and may ultimately break through areas of weakness in the Earth's crust. When this happens, an eruption occurs and the molten rock may pour from the vent as non-explosive lava or it may shoot violently into the air as dense clouds of lava fragments. Larger fragments fall back around the vent, and the accumulations of the fragments may move down slopes as ash flows under the force of gravity. Some of the finer ejected materials may be carried by the wind only to fall to the ground many miles away. The finest ash particles may be ejected miles into the atmosphere and can then be carried around the world by stratospheric winds before settling out. Magma is the molten rock below the surface of the Earth that rises up through the volcanic vents. The term lava is used for magma that has erupted through the vents and originates many miles beneath the ground. Ascending magma commonly contains some crystals, fragments of surrounding unmelted rocks, and dissolved gases, but it is primarily a liquid compound composed of oxygen, silicon, aluminum, iron, magnesium, calcium, potassium, titanium, and manganese. Magmas also contain many other chemical elements in trace quantities. Upon cooling, magma forms igneous or magmatic rock. It is important for people who are living close to volcanoes to leave their homes and go to a safer area soon after they hear about volcanic eruption/activity.

## Flooding

Flooding kills people and destroys homes in many parts of the United States. About 125 people die of flooding in the US every year. Property damage runs into billions of dollars and of late these figures have risen consistently. Flooding is caused by a number of reasons, including flooding during hurricanes due to heavy rains, coastal flooding (sea/ocean waves), inland flooding from snow melting or swollen rivers, flooding from failure of dams, impoundments or other water regulatory systems, flash floods from sudden large downpours and also from ice jams (melting of ice due to heat). Every year the US government spends billions of dollars assisting its citizens. People need to leave their homes in the event of flooding and also insure their properties against flood and other natural disasters.

## Snow and ice

Snow and ice are well known hazards to those living in mountainous areas (regions north of 35 degrees latitude). During the winter months, prolonged power outages, automobile accidents, transportation delays, damage to buildings, and dangerous walkways are often attributed to snow and ice. Although slippery surfaces are often the primary cause of such problems, reduced visibility is responsible for many accidents. Snow can be warm causing wet and slushy conditions, or cold, creating dry and powdery conditions. The latter leads to blizzards when mixed with high winds. Blizzards can reduce the visibility to zero. Drifting can block roadways, airport runways, and even bury buildings. Snow and ice are often associated with low wind chills, which are dangerous to exposed skin.

Snow and ice reduce visibility and when they accumulate on the surface, they reduce traction and put strain on power lines, roofs and other structures. The most important part of dealing with snow and ice is forecasting and issuing warnings for the purpose of community preparation. Communities can prepare for winter storms by stocking sand and salt to improve road conditions, advising people to use caution when leaving their homes and to stock plenty of food before a storm in case they are confined to their homes.

## Extreme heat

Extreme heat is very dangerous to people in tropical and sub-tropical areas. Sometimes even in places which are colder in winter, summer can be very hot and heat waves can sweep through. From 1979 to 2002, excessive heat exposure caused 8,966 deaths in the US. During this period, more people in the country died from extreme heat than any other natural hazard. The elderly, children, and people with certain medical conditions (such as heart disease) are at greater risk. Even healthy individuals can become victims to heat if they exercise strenuously in extreme heat. It is often referred to as "Heat Stroke" in the tropics. This can be avoided by taking some precautions. The most important of which is to remain indoors, install air conditioning in homes, and increase the intake of water and fruit juices. It is important from a humanitarian point to check on neighbors who may be susceptible.

### SKILL 16.5    Explain the role of the National Weather Service in issuing weather alerts.

The National Weather Service (NWS) is one of six scientific agencies that comprise the National Oceanic and Atmospheric Administration (NOAA) of the United States Government. The NWS is responsible for the collection of weather, hydrologic, and climate data at national and regional centers, and for the provision of forecasts and warnings for the United States based on this data.

## Collection of Data

The NWS gathers weather related information using surface, marine, and upper air observations.

Surface observations are organized by the Automated Surface Observing System (ASOS), which records hourly and special weather observations at transmission sites (usually located near airport runways). Conditions recorded at these stations include sky condition, visibility, present weather, obstruction to vision, pressure, temperature, dew point, wind direction and speed, and precipitation accumulation.

Large bodies of water have an important impact on the weather. For this reason, marine observation data and the output of numerical water models are crucial NWS forecasts. To record marine conditions, the NWS's National Data Buoy Center (NDBC) employs a system of 90 buoys and 60 land-based coastal observing systems. Hourly, these stations report wind speed, direction and gust, barometric pressure, and air temperature. Additionally, some stations report conductivity, water current, sea surface temperature, and wave height and period.

Upper air observations are recorded by the NWS's 92 radiosonde locations in North America and 10 sites in the Caribbean. A **radiosonde** consists of a small, expendable instrument package suspended below a 2 meter-wide balloon filled with hydrogen or helium that is released daily at 00 UTC and 12 UTC. Sensors on the radiosonde measure and transmit wind speed and direction, pressure, temperature, and relative humidity recordings. Data from the flight of a radiosonde are plotted on a Skew-T or Stuve diagram for analysis.

## Using Data to Issue Weather Alerts

The NWS has developed a multi-tiered set of terms to provide observation-based weather forecasts and alerts to the United States public.

Outlook

This alert is issued daily to address potentially hazardous weather or hydrologic events that may occur in the next seven days. A hazardous weather outlook is intended to provide information to those who may need a considerable amount of time to prepare for future weather events, such severe thunderstorms, heavy rains, flooding, winter weather, extremes of heat or cold, etc. This type of alert may also be issued on an event-driven basis, such as the Flood Potential Outlook and Severe Weather Outlook.

### Watch

This alert is issued when the risk of a hazardous weather or hydrologic event has increased significantly, but its occurrence, location, or timing remains uncertain. Similar to outlooks, watches are intended to provide information to those who may need time to prepare for a weather event. A weather watch suggests that hazardous weather is possible, but not certain, and that people should establish a plan of action in case the weather event occurs. Watches also recommend that people listen for later information and warnings, especially when planning travel or outdoor activities.

### Warning

A warning is issued when a hazardous weather or hydrologic event is occurring, imminent, or likely. Warnings are therefore more serious than watches. This type of alert means that weather conditions pose a threat to life or property, and that people in the path of the storm need to take protective action.

### Advisory

Advisories are similar to warnings in that they are issued when a hazardous weather or hydrologic event is occurring, imminent, or likely. However, advisories indicate that conditions are less serious than those of weather warnings.

## TEACHER CERTIFICATION STUDY GUIDE

**COMPETENCY 17.0**     **Understand the locations and characteristics of the Earth's major climatic regions and analyze factors that affect local climate and the relationship between weather and climate.**

**SKILL 17.1**     **Infer the climatic zone in which a given area is located based on temperature and precipitation data.**

The Köppen climate system explains that the origin of the climate is based on the average monthly and yearly temperatures, location of the landmass, precipitation rates, and seasonality of the precipitation. When Köppen, a climatologist, created his classification system, he determined that one of the best indicators for climate is native plant life. He created the Köppen climate system map with native vegetation in mind.

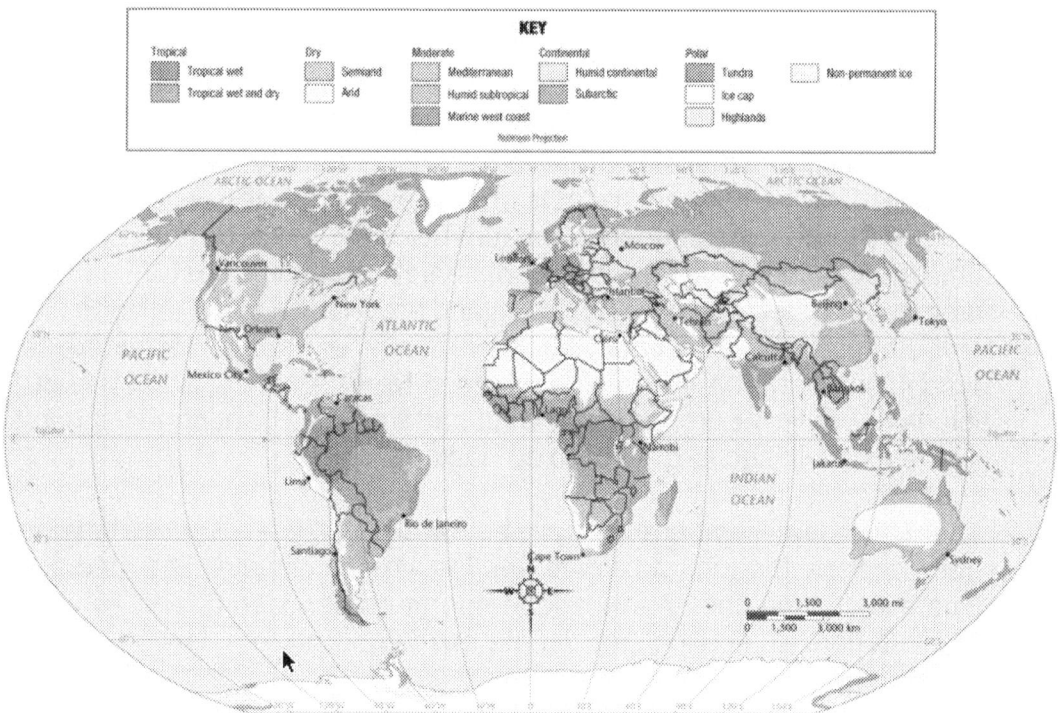

## Types of Climate

1.) **Tropical/Megathermal Climate**: Characterized by a constant high temperature over 18°C or 64.4°F. There are three subcategories: tropical rain forest, tropical monsoon, and tropical Savanna. The tropical rain forest climate has more than 60 mm of rain each month of the year. There are no season changes. The tropical monsoon climate will have more than 100 mm of rain total in the year, but may have some months that fall below 60 mm of rain due to change in wind direction as the seasons change. The tropical savanna climates have a prominent dry season and have less than 100 mm of rain each year.

2.) **Arid and Semiarid Climate:** The precipitation of the region is less than the evapotranspiration of the region.

3.) **Temperate/Mesothermal Climate:** The average temperature is above 10°C for the warmest months and between -3°C and 18°C in the coldest months. There are four subcategories: Mediterranean climate, humid subtropical, maritime temperate, and maritime subarctic. The Mediterranean climate is found on the western sides of continents. It has moderate temperatures and experiences a polar front in the winter and a tropical front in the summer. Summers are usually hot and dry. The humid subtropical climate is found on the interior of the continents or on the east coast of the continents. The summers are usually very humid due to the trade winds bringing moisture to the region. The maritime temperate climate experiences a polar front all year. It is usually found on the continents between 45° and 55° latitude. The weather is usually overcast year round. Finally, the maritime subarctic climate is closer to the poles than the maritime temperate climates and is usually limited to thin strips of land or islands off the western coast of the continents.

4.) **Continental/Microthermal Climate:** Characterized by temperatures above 10°C in the summer months and below -3°C in the winter months. There are three subcategories: hot summer continental climate, warm summer continental climate, and continental subarctic climate. The hot summer continental climate occurs inland around 30° to 40° latitude. They can be affected by monsoons. The warm summer continental climate is found inland between 40° to 50° latitude in North America, and up to 60° latitude in Eastern Europe due to wind patterns. The continental subarctic climate exists inland in the 50° to 60° latitude.

5.) **Polar Climates:** Temperatures are below 10°C all year. There are two subcategories: the tundra climate and the ice cap climate. The tundra climate is dry and has an average temperature between 0° and 10°C in the warmest months. The ice cap climate has temperature below 0°C year round.

## Variation in Climate

Climate varies over time. The diurnal cycle is how the weather has changed in the last 24 hours. Over a year, regions experience changes in seasons. Climate of a region may vary due to changes in atmospheric dust from large dust storms and volcanic eruptions. Over the course of 10 years, changes in climate may take place due to the el Niño and la Niña cycles in Earth's oceans. Climate change over the course of 100 years can be attributed to solar variability and changes within ocean temperature. Over the last 100 years, significant climate changes have been noted in some regions due to deforestation and increased carbon dioxide output. Climate variability over 1000 years can be linked to thermocline circulation in the oceans and changes in the carbon cycle. Paleoclimatology indicates that climate variation on the 100,000 year time frame can be attributed to the Milankovich cycles, solar variability, precession, and orbit eccentricity. By studying variations in climate, paleoclimatologists can determine how current changes in climate are related to long term trends. This can help scientists to further understand humanity's effect on climate change.

**SKILL 17.2** **Identify the factors that affect the climate in a given region (e.g., insolation, wind patterns, topography).**

The weather in a region is called the climate of that region. Unlike the weather, which consists of hourly and daily changes in the atmosphere over a region, climate is the average of all weather conditions in a region over a period of time. Many factors are used to determine the climate of a region including temperature and precipitation. Climate varies from one place to another because of the unequal heating of the Earth's surface. This varied heating of the surface is the result of the unequal distribution of landmasses, oceans, and polar ice caps.

## TOPOGRAPHY

The climate, topology, and atmospheric processes all combine to trap pollutants. **Inversion**: an atmospheric condition where the air temperature rises. Inversion is driven by temperature. Warm air rises only so far. If it encounters an atmospheric pressure cell of colder air it will stall. Colder surface air attempting to rise won't go past the higher altitude warm air. This traps pockets of pollutants in the air.

Topology also acts to trap pollutants. Mountains and valleys are the key locations. In valleys, the normal flow of air is downward during the daytime, and at nighttime, from the valley floor up the sides of the mountains. However, during winter months the airflow pattern can reverse and become a down flow all the time. The down-flowing cold air traps the warm air layer filled with pollutants in the valley.

### The Radiant Heat Energy Balance (The Radiation Budget)

**Radiation Budget**: the balanced exchange cycle of radiation absorbed and released by the Earth's surface, water, and atmosphere.

The Sun drives a radiation exchange interaction with the Earth every day as the Sun's radiated energy enters the Earth's sphere of influence. The 19% absorbed by the $CO_2$, dust, clouds and water vapor in the atmosphere traps the heat and holds it to moderate the temperature (especially at night). In the desert, there is little water vapor present to trap the heat. That's why it's so cold at night. The interaction between the Earth's upper atmosphere, waters, and land keeps the world at a moderate ambient temperature of approximately 55 degrees F.

### Solar Radiation (Insolation)

This distribution of solar energy is called **insolation**. Solar radiation isn't distributed evenly across the Earth because of the Earth's curvature, axial tilt, and orbit. This results in uneven heating of the atmosphere, and is why the temperature is warmer at the equator and colder at the poles.

Because of the curvature and tilt, the energy striking the polar areas is spread over a larger area. At the equator it is more concentrated. The same amount of energy is striking the atmosphere, but it is striking a larger or smaller area. In effect this dilutes the energy received by a particular area.

The effect of insolation is very important to life on Earth. The absence of solar radiation would cause the creation of very cold air masses and the thermal blanket of the atmosphere would not have heat to hold and reradiate. In short order, the world would become an icy rock.

### Air as an Insulator

Air in the atmosphere is a good insulator, analogous to a goose-down blanket.

**Convection Cells**: The ground absorbs energy during the morning hours and re-radiates the absorbed energy in a narrow band near the surface. This heats the air near the surface. The air near the surface absorbs the energy and expands. Convection is the primary means of transferring heat to the air.

As the molecules move faster the air expands, becomes less dense, and rises. The air continues to rise, but begins to cool, again becoming denser, and it starts to sink back to the surface. This cycle of rising, warming air and sinking, cooling air is called a convection cell.

**Parcel of Air (Air Mass):** a large area of air, which assumes a characteristic temperature, pressure, and humidity from sitting over a landmass. Horizontally, the parcel has relatively uniform temperature, pressure, and humidity. Vertically, the parcel may have widely differing temperatures, pressures, and humidity. When you think of air parcels remember to think in the horizontal plane.

**SKILL 17.3**      **Describe the relationship between the climate of a region and its weather.**

**Climate**: the characteristic weather of a region over long periods of time.

**Weather**: the state of the atmosphere at a particular time and place.

The Hydrologic Cycle is greatly influenced by the climate. Warm air rises and holds more water than cold air. Cold air sinks and holds less water than warm air. This temperature differential balances the water cycle. Much water is locked up in glacial ice. As the climate changes, glaciers melt or freeze, changing the volume of water available around the world.

The various biomes are characterized not only by the soil types and organisms present in them, but by their unique weather patterns. Indeed, the climate largely determines what organisms evolved to populate each of these ecosystems. Along with temperature profiles, precipitation patterns are the key component of the climate of an ecosystem. One of the most dramatic examples of a weather event's importance to an ecosystem is the annual rainfall on some arid grasslands and deserts. Though the plants and animals that live in these grasslands are adapted to dry conditions, they rely on the yearly rains to maintain their life cycles. Similarly, many ecosystems experience regular seasonal changes that provide cues to the organisms therein. The seasonal changes in temperature and precipitation may trigger the plants to sprout, drop seed pods, or become dormant and similarly signal animals to breed, eat more or less food, or enter hibernation. Even seemingly catastrophic events such as fire or flood are often part of the normal cycle of life in some ecosystems. Certain species have evolved mechanisms that allow them to survive these events, while for others, it is a time of renewal that removes old individuals and triggers growth of a new generation. This recovery is part of the natural phenomenon of succession, which is a normal and necessary process.

It should be clear then, that changes in climate would have significant effects on the species within an ecosystem. Because each biome's organisms are uniquely adapted to the conditions of that biome, they typically will not fare well when climatic or other changes are introduced. Changes in precipitation, annual temperature profiles, or availability of nutrients may shorten or even eliminate the growing season of an area. This, in turn, creates less food for the heterotrophic animals and may cause the death of organisms. Additionally, climatic changes may cause the loss of aquatic or arboreal habitats, thus endangering the species that rely upon them. Finally, if significant enough temperature changes are introduced, certain species may be unable to maintain their metabolism at the new temperature. Of course, some organisms will be able to survive in new conditions and even adapt to thrive.

**COMPETENCY 18.0** Understand the effects of human activities and natural processes on the atmosphere, theories about the long-range effects of human activities on global climate, and methods of controlling and minimizing those effects.

**SKILL 18.1** Identify common air pollutants and their sources and effects.

### Types of Pollutants

**Primary Pollutants**: chemicals released to the air in a harmful form.
Example: Smokestack exhaust is harmful if not properly scrubbed before release.

**Secondary Pollutants**: chemicals that are not harmful as released, but are modified in the atmosphere and become harmful.
Example: Exhaust emissions of Nitrous Oxide from automobiles. When exposed to sunlight, it becomes harmful.

**Fugitive Emissions**: emissions released by mistake.
Example: Holes in automobile tail pipes, mufflers, catalytic converters, and failed industrial scrubber units.

### Sources of Outdoor Pollutants

Although there are natural pollutants, the primary sources of outdoor air pollution are pollutants created by humans. Transportation sources alone account for almost 50% of all outdoor air pollution.

The U.S. **Environmental Protection Agency** (EPA) recognizes seven substances as being harmful to the environment. These are:

Sulphur Compounds: Some are naturally released. The primary natural sources are volcanoes. The man-made sources are primarily from the combustion of coal and oil products. When released, the sulfur combines with the atmosphere and changes to an oxidizer, Sulfur Dioxide ($SO_2$). Further chemical modifications transform this into Sulfate, and when combined with water, it changes to Sulfuric Acid ($SO_4$). This acid precipitation is known as Acid Rain. The solution to the problem is to use low sulfur content coal. Since the EPA mandated its use, there has been an 11% drop in the amount of acid rain.

Nitrogen Compounds: 5% of all Nitrogen compounds occur naturally: bacterial decay, rice paddies, wetlands, and swamp decay. 95% of the problem is caused by man-made, high temperature combustion. Auto exhaust is the primary offender. The nitrogen compounds oxidize and combine with $H_2O$ (water) to form into $SO_4$ (acid rain).

Carbon Oxides: Carbon Monoxide (CO) is very deadly because it replaces oxygen molecules in the blood, causing Carbon Monoxide poisoning. Carbon Monoxide is produced by combustion of oil and gas, with auto exhaust as the primary offender.

Particulates: These are Aerosols: small bits of solids suspended in $H_2O$ (water) drops. Aerosols block solar radiation and make the air look dirty. The source of the particulate (dust, soot, ash) can be either man-made or natural. 40% comes from industrial processes. 17% comes from vehicle emissions. 30% comes from natural sources: salt spray, dust, and volcanoes. 13% comes from all other miscellaneous sources. The most dangerous of the particulates are <2.5 microns in size. These microscopic sized particles cause lung problems. Some sources are asbestos, cigarette smoke, and coal dust. Example: Long exposure to coal dust can cause "Black Lung" disease. The sulfur in dust particulates literally burns acid holes in the lungs.

Metals: Metal substances are the byproducts of a variety of manufacturing processes. Some sources are sandblasting, leaded gasoline and leaded paint. Example: Since the EPA banned the use of lead in gasoline and paint there has been a major drop in lead accumulation in the environment. The major problem with metals is that almost all of them are carcinogenic, and because the metals accumulate in the fatty tissues of the body, they also cause a long-term problem with nervous system disruption. Example: The haberdashery (hat maker) trade of the past had a high rate of craftsmen going insane. The reason was that the trade used to use a large amount of mercury in the crafting process. The fumes from the mercury were breathed by the workers, and eventually caused severe nervous system disruption. Thus comes the old phrase "Mad as a Hatter".

Photochemical Oxidants: The oxidation process can change normally harmless chemicals into more dangerous substances. Example: Oxygen ($O_2$), when exposed to ultra-violet radiation (UV), becomes Ozone ($O_3$). Although Ozone is necessary in the upper atmosphere, when the human body is exposed to it, it causes lung damage. Example: Nitrogen compounds under certain circumstances can transform into $NO_3$. This compound is a major component of smog that irritates the soft tissues and mucous membranes.

Volatile Organic Compounds (VOCs): These are organic chemical gases that occur both naturally and through man-made processes.

Natural sources: Plants, wetlands, rice paddies, and ruminant animals release Methane. Conifer trees release Terpene, which forms a haze in the atmosphere. It is what forms the haze over the Great Smoky Mountains and the Azure Mountains.

Man-made Sources: Any synthetic organic chemical (such as Benzene, Toluene, and Formaldehyde). All hydrocarbons emissions such as gasoline fumes.

Controlling Air Pollution Outdoors

**Cyclone Collectors**: Utilizes an electrostatic precipitator. Small particulates are electronically attracted to a charged filter screen and destroyed.

**Scrubber**: Used to control $SO_2$ (Sulfur Dioxide) emissions. Limestone is crushed and mixed into a slurry. The slurry mixture chemically removes 95% of the $SO_2$ emissions. The use of scrubber units is a relatively inexpensive procedure. However, the United States is the only country to consistently use this method. In 1987, only 40% of German industries used this method, compared to 95% of U.S. industry.

**Afterburners** and **Converters**: Nitrogen compounds in auto exhaust can be removed by the use of afterburners. Afterburners rely on another stage in the combustion process to re-channel the exhaust back to the combustion source and re-burn it. Catalytic converters use platinum to act as a catalyst to remove $CO_2$ and CO compounds.

**Indoor Air Pollutants**

Indoor air pollution is on the rise due to urban lifestyles. New construction techniques cause better sealing of buildings, trapping air inside. If the air filtration system isn't sufficient or properly maintained, then the occupants end up re-breathing polluted air.

Carbon Monoxide: Produced by gas or oil burning stoves, water heaters, furnaces, fireplaces, etc., the gas is colorless, odorless, and deadly. Preventive safety measures include ensuring proper ventilation and use of Carbon Monoxide (CO) detectors.

PAH (Polycyclic Aromatic Hydrocarbons): This gas is released by new carpeting, flooring, fabrics, and padding made from polycarbonates. Long-term exposure to PAH causes eye tearing and aggravates asthma and bronchitis.

Radon: A naturally occurring, radioactive gas. The source of the emission is the natural decay of radioactive elements from granite, uranium and other metamorphic rocks with contained radioactive elements. As the rock material weathers or breaks down, the gas seeps into the foundations of buildings and homes. The gas then can be re-circulated through the air system. The gas is a carcinogenic and in areas where this type of material forms the bedrock, the use of Radon detectors is advised.

Formaldehyde: This chemical is used extensively in the processing of fabric. Fumes from the trace amounts left on the fabric cause teary eyes and skin rashes.

Asbestos: A naturally occurring mineral, asbestos was once extensively used as a fire retardant in homes, buildings, schools, and as a lining for automobile brake shoes. In its natural state (as a solid) it is harmless. However, if the fibers become airborne, they can collect in the human lungs and eventually cause asbestosis. Exposure to airborne particulate asbestos fibers can be deadly in the long term. In the short term, it can cause severe respiratory problems. Its use is now banned by the EPA, and the rip out and replacement process is very time consuming and expensive.

Tobacco Smoke: The EPA has identified this as a major carcinogen. The EPA has also identified second hand tobacco smoke as a problem. The effects of tobacco smoke are lung cancer, emphysema, heart disease, asthma, bronchitis, and Brittle Lung Disease (lungs lose their elastic capacity). EPA data is the major source material cited in the enactment of antismoking legislation.

**SKILL 18.2** **Describe pollutants and atmospheric chemical reactions involving these pollutants.**

**Acid deposition** can include anything acidic: particulates, fog, rain, dew, and aerosols. Acidic deposition affects the pH balance of the environment. Normal balance is 5.5 to 6.5. Water ($H_2O$) and carbon dioxide ($CO_2$) can chemically combine to become Carbonic Acid ($H_2CO_3$). Sulfur Dioxide ($SO_2$) particles can chemically combine with water ($H_2O$) to form Sulfuric Acid ($H_2SO_4$).

Direct effects of acidic deposition

- Kills trees: Over 6 million trees in Scandinavia have died over the last decade.
- Destroys the ecosystem of lakes: With a pH balance of 5.0, the small fry and eggs die, and the reproduction cycle is interrupted. With a pH balance of < 5.0, the adult fish die. Some fish will survive, but it depends on the hardiness of the species. Liming the lake can counteract the effect. However, this is expensive and only a short-term solution.
- Monuments and buildings are slowly being destroyed. The acid is rapidly weathering the carbonates in limestone structures or anything with a calcium carbonate composition. Famous statues and architectural features are slowly being eaten away. Heavy, toxic metals are being released into groundwater sources.

**SKILL 18.3    Describe factors that affect local air pollutant concentrations (e.g., population density).**

Normal, daily human activity can generate emissions that produce air pollution. These activities may include transportation, electricity use, solvent use, adhesives, paints, and home heating and cooling. As population density increases at particular geographic locations, the frequency of such activities increases, affecting local air pollution. Human use of products that release CFC's into the atmosphere contributes to increased air pollution. In areas of higher population density (urban areas) the two largest sources of air pollutants are transportation and fuel combustion.

Transportation

Automobiles produce high levels of carbon monoxides and nitrogen oxides, both of which are greenhouse gases. As populations rise, vehicle use also rises. Recent studies performed for the Federal Highway Administration found that, generally, traffic volumes tend to rise at least 80% of the rate of population density increase. This rise is explained by the increasing number of people who own and use cars in a growing city. In addition to the increased number of cars used by a growing population is the problem of increased traffic. As the number of cars increases in a particular geographic location due to growing population density, the traffic slows down and becomes subject to more stop-and-go operation. This situation further increases the amount of time an automobile is in use, worsening air pollution.

Fuel Combustion

Another source of air pollution is fuel combustion by stationary sources including residential, commercial, and industrial heating and cooling and coal-burning power plants. Fuel combustion is the dominant source of sulfur dioxide air pollution. As population density increases, the number of people requiring and using energy increases. More power is needed to construct new homes and industries, to heat and cool a larger number of homes and buildings and to generate more food and commercial products. Energy used in all of these situations may be produced by the burning of fossil fuels, thus increasing the amount of greenhouse gases released into the atmosphere.

**SKILL 18.4   Analyze the theory of global warming due to increased levels of atmospheric carbon dioxide from the burning of fossil fuels.**

One of the major environmental issues of today is the controversy over **Global Climate** change. Many scientists forcefully contend that we are experiencing a period of global warming, and that our dependence on modern technology is the major reason behind the warming. Not surprisingly, other scientists equally oppose this theory. Much of the controversy depends on the source of information as to whether humankind is at fault.

### Is the Earth Warming or Cooling?

In the 1970's the scientific issue was whether there was going to be another ice age. However, as we approached the new millennium, the issue shifted to concern about warming. The controversy is based on interpretations of the historical records. Temperatures have been recorded based upon four temperature readings per day from each collection location. The collected readings are sent to the World Metrological Organization where the readings are correlated and averaged. That data indicates a rise in average temperature. However, the temperature readings only have been collected for the past 100 years.

### Problems with the Collected Data

- **How were the records constructed?**

Weather stations use an averaging method. In 1945 the World Meteorological Organization dictated the method of collection, type of shelter, height off of the ground, equipment used, etc. in order to normalize the temperature readings world-wide. This make previously collected records suspect as to the means through which they were collected.

- **The locations of weather stations.**

The big problem is where to place the weather collection stations. In example, where is the ideal place to collect ambient temperature? In reality, the stations were placed at locations where they could be monitored roughly 24 hours a day. Most of the stations are placed at airports. However, because of the expanse of heat absorbing building materials (i.e. concrete and asphalt) airports are usually hotter than surrounding areas. Stations located within other **Urban Heat Islands** (i.e. the cities) are also subject to same problem.

- **How to collect and analyze correct data?**

The most accurate method to measure global temperature is to use space satellite technology. Since 1979, the GOES Satellite has collected temperature readings. Instead of sampling only the cities, the GOES samples all areas in blocks of data, including the oceans. Data taken during the period of 1979-1998 showed no net warming. However, from a scientific point of view this is not a long enough period of sampling to be fully conclusive.

**Has the Climate Always Been the Same?**

**Paleoclimate**: the climate of the past. The paleoclimate is studied by CLIMAP, an organization that is focused on past climatological situations. Both Fine and Gross methods are used to determine the past climate.

**Gross Methods**

The study of arboreal forests, cactus, coral, etc, gives a rough idea of the past climate based on the fossil record. It can also show the presence of species that favored warm or cold climates, providing clues as to what the climate in a location used to be.

The study of historical records also provides important information. People write about what is going on in their lives. In-depth research in old texts, letters, etc., can provide clues. 100 years ago grapes grew in England. This is not so today.

As the climate changes, people comment on the change, or popular culture reflects the fears and issues caused by the changes. Stories of trolls, ogres, and dwarfs were probably reflecting medical conditions caused by the little ice ages. The little ice ages caused a shortage of food. This led to dietary deficiencies and malnutrition. The malnutrition manifests itself as stunted growth, gnarled joints, etc, which gave rise to fairy tales of deformed creatures.

Studying what the topographical records indicate is also useful. Example: Large Rivers used to run through the Sahara Desert. The presence or absence of ancient seabeds, riverbeds, etc. may provide clues of past climate changes.

**Fine Methods**

Fine methods focus on improved scientific methods of taking past temperatures.

**Oxygen Isotope Ratio Measurements**: all living things utilizing oxygen retain a percentage of oxygen within their cellular structure. The change in the ratio of $O_{18}$ and $O_{16}$ provides clues as to the type of climate they lived in. Carbonate shells and coral remains on the ocean floor are key fossils. More $O_{18}$ indicates a colder climate. More $O_{16}$ indicates a warmer climate. The oxygen trapped within the $CaCO_3$ bones and shells gives a record of the air at the time the creature lived.

Air Bubbles trapped within Glacial Ice: Analysis of the ice sheets in Antarctica, Greenland and Siberia provide clues. As you drill down the layers of ice provide a record of changes, much like tree ring dating techniques. Analysis of the air bubbles is very accurate. It provides data on temperature, gaseous content concentrations, etc. The analytical methods used can be applied to samples as far back as 160,000 years in the past.

**Climate Trends**

What we know about the climate in the distant past is based entirely on Gross methods (lithology and the fossil records). There were long periods of warm, stable climate punctuated by climatic changes.

**Climatic Disruptions**: a period of glaciation followed by a warmer stable period that is interrupted by interglacial periods of 5 million years.

The last ice age ended approximately 12,000 years ago. Ice covered all of Canada and most of the upper United States. It also covered large portions of Europe.

Our knowledge of the climate in the recent past is more complete. The information is based upon fine-methods (glacial ice air bubble analysis and oxygen isotope studies of carbonate remains) and indicate that climate change is not new.

Although it is yet unclear which effect leads the other, the correlation between temperature change and Carbon Dioxide ($CO_2$) concentrations show a direct inverse relation between the two.

**EARTH SCIENCE**

The data analysis indicates we are warmer today than 160,000 YBP. However, since the data analysis-starting point was in the middle of an ice age, the sample may be too small and the analysis basis somewhat suspect.

**Reasons for Climate Change**

**The Greenhouse Effect**

The Greenhouse Effect is predicated on a rise in the trapped gases in the atmosphere.

$CO_2$, Methane and Water Vapor all absorb reflected heat. $CO_2$ makes up 0.03% of the atmosphere. If the $CO_2$ level rises, more reflected heat is trapped, and the balance of the **Radiation Budget** is upset.

When greenhouse gases and heat build up, the Earth's surface and atmospheric temperatures rise. The theory contends that if we cut the amount of rising $CO_2$ in the atmosphere, temperatures will fall.

**Sources of greenhouse gases**

Gasoline burning engines and Aerosols ($CO_2$)
Vegetation & Microwave devices ($CO_2$)
Rice Paddies & Ruminant animals (Methane)

$CO_2$ levels have risen this century, with the most dramatic rise since the 1960's. However, the data is not conclusive when compared to the past. The only thing certain is that there is a one-to-one correlation between temperature and $CO_2$.

Are the oceans a $CO_2$ sink? The possibility of the oceans acting as a major source of $CO_2$ release cannot be discounted. Warm liquids hold less gas. As liquids cool, the gas goes into solution. Therefore, a change in the ocean temperatures either releases or holds more gases.

**Milankovitch Cycles**

This theory proposes that the axial tilt and wobble of the Earth's orbit is responsible for warming or cooling.

**Precession**: the wobble of the Earth on its axis.

**Obliquity**: the tilt of the Earth's axis. This ranges between 22.5 to 25.5 degrees. Today the tilt is 23.5 degrees.

**Eccentricity**: the shape of an orbit. The Earth's orbital shape changes periodically from elliptical to circular.

The theory is based on the Earth's relationship to the Sun. If there is a change in the tilt and orbit, then the amount of incoming solar radiation will change over time. The tilt of the axis is the key determinant in directing where most of the solar radiation strikes. It causes either longer or shorter summers and winters. A shorter summer increases the snow pack in the mountains. As this builds over time, it causes an interglacial or glacial period. The cycle is between 18,000 to 100,000 years. With the exception of one period, ice core $O_{18}$ ratio data correlates with this cycle, and also shows a possible correlation between the Milankovitch Cycles and the oceans as a $CO_2$ sink.

**Other Causal Ideas for Climate Change**

Tectonic Movement
The oceans circulate the distributed heat while mountains block the circulation. This blocked circulation causes a temperature rise. The problem with this theory is that it doesn't match the ice core data.

Volcanic Activity
It's known that there is a relationship between volcanic activity and temperature change. The amount of aerosols increases the reflectivity, thereby decreasing the temperature. However, these are generally short-term, more localized effects.

Increased Cloud Cover
Increased evaporation creates more clouds. An increase in cloud cover causes a decrease in temperature. We need more data as no long-term studies on this theory have been conducted.

**The Effects of a Warmer World**

**Melting of the Polar Ice Caps**
This would cause sea level rises of 50 to 150 ft.
Much of the present day coastal shorelines would be underwater.
New York City would be gone.
Most of Florida would be submerged.
Major California cities inundated with water.

**Shifts in the Rain Belts**
World wide crop failures with resulting famine.
The mid-west United States would have a climate like present day Arizona.
Canada would become a major agricultural source

**Warming of the Oceans**
Exceptionally strong hurricanes and typhoons would ensue.

## TEACHER CERTIFICATION STUDY GUIDE

**SUBAREA IV.**                        **GEOLOGY AND OCEANOGRAPHY**

**COMPETENCY 19.0**     Understand the processes of mineral and rock formation, the characteristics of different types of minerals and rocks, and the methods used to identify and classify them.

**SKILL 19.1**     Use classification schemes (e.g., Mohs' scale of hardness, crystal form, chemical composition) to identify common rock-forming minerals.

### Minerals

A mineral is a naturally occurring homogenous element or compound of homogenous elements with a definite chemical composition (within limits), and a highly ordered atomic arrangement (crystalline structure).

Some minerals are single elements. An example of this is a diamond, which is solely Carbon (C). Some minerals are simple combinations of atoms. Examples of this are Halite (NaCl) and Pyrite ($FeS_2$). Other minerals are complex combinations of atoms. An example of this is Tourmaline whose formula is $Na(MgFe)_3AlO(BO_3)(Si_2O_{18})(OH)_4$.

The $(Mg, Fe)_x$ portion of the formula indicates that the Magnesium and Iron may combine in any combination of 3 atoms as long as both elements are present in the final composition. This type of formula variance is why the definition of a mineral states that the chemical composition must be within limits. This variance in formula also causes variance in coloration of the same mineral. Variance in color is caused by small impurities.

Crystalline structure, form or shape (not chemical composition) is the key defining factor in identifying minerals. Crystalline structure is defined as the most efficient arrangement of atoms that form a crystal shape.

Minerals are arranged in specific patterns based on their elemental composition. For example, Halite (NaCl) forms into cube shapes, while a diamond's structure is a complex latticework of diamond shapes.

**Crystal Habit**: the shape of the mineral. Some minerals possess distinctive shapes.

**Cleavage**: how the mineral breaks under pressure. Most minerals have a tendency to break in a preferred direction along smooth surfaces. Where the atoms connect, it forms a weak point. In example, Mica is resistant to breaking but peels quite easily. Not all minerals have cleavage. Instead, some have fractures (they shatter like glass).

**EARTH SCIENCE**

**Hardness**: how hard the mineral is. Hardness is based upon the arrangement of atoms within the crystalline structure. Hardness is graded from 1 to 10 using Mohs Scale of Hardness. General classification is acceptable based on the scratch test.

**Mohs Scale of Hardness**
1. Talc
2. Gypsum
3. Calcite
4. Fluorite
5. Apattte
6. Orthoclase (Feldspar)
7. Quartz
8. Topaz
9. Corundum
10. Diamond

**Soft**: Able to be scratched with a fingernail.
**Hard:** Able to scratch glass with the mineral.
**Medium:** not able to be scratched with a fingernail, nor able to scratch glass with the mineral.

**Specific Gravity**: the ratio of the mineral's weight to water. Because the weight of the mineral is based upon the arrangement of the atoms, minerals will vary in specific gravity (e.g., water = 1, but rock = 2.65).

**Color**: the color of the mineral in solid form. Although some minerals, such as Sulfur, are distinctly colored- always some shade of yellow- it's relatively common to find color variations within the same mineral. As a result, identifying minerals using color as the sole basis for identification can easily lead to misidentification.

**Streak**: the color of the mineral in powdered form. Some minerals leave a distinctive color streak when the mineral is scratched across a Streak Plate (a piece of unglazed porcelain).

**Additional (Secondary) identifying Properties**

**Luster**: the surface appearance of the mineral.
Examples of luster are Pearly, Waxy, Shiny, Dull, Earthy and Glassy.

**Magnetism**: inherent magnetic qualities.
Example: Magtletite=yes, Quartz=no.

**Fluorescence**: Some minerals glow under a black light.
Example: Lapis Lazuli.

**Reaction to Acid**: Some minerals have a distinctive reaction when exposed to acids. For example: Any mineral with calcium carbonate ($CaCO_3$) will fizz when diluted hydrochloric acid (HCl) is dropped on it.

**Striations**: distinctive marks on the surface of the mineral. These marks are usually parallel lines on the mineral's surface. In example, Feldspar is often heavily striated.

**Taste**: Some minerals have a distinctive taste.
Example: Halite (NaCl) (more commonly known as table salt).

Minerals are divided into three common types: Silicates, Orebites, and Carbonates.

**Silicates**

Silicates are the most common type, making up 90% of all minerals.
The Silica tetrahedron of Oxygen ($O_2$) and Silicon ($Si_4$) is the basic crystalline structure of a silicate.

The silicates are subdivided into two divisions: pure silica (Quartz) and silica combined with other elements (Ferromags and non-Ferromags). Both Ferromags and non-Ferromags are silicates. The difference is that one has iron and magnesium in it and the other doesn't.

Any silicate mineral that is a Ferromag has iron (Fe) and magnesium (Mg) in its formula. Ferromags are dark in color and dense in mass. An example of a ferromag is Hornblende.

Non-Ferromags are silicate minerals with chemicals other than iron or magnesium in their chemical composition. They are light in both color and mass density. An example of a nonferromag is Feldspar.

If examined in a thin, wafer-like section, their crystalline structure clearly shows that the silica tetrahedrons form chains in distinctive patterns that are indicative of the grouping of the silicate.

**Single-Chain Structure**

The corners of the tetrahedron coincide with the centers of the oxygen ions. When other elements combine with the silicate tetrahedron, they do so at the junction points of the chains. In the case of Biotite (Mica), it has iron combined in a thin sheet over the entire surface of the tetrahedron.

Silicate Structures include Double-Chain (Amphibole Group: "Hornblende") and Sheet Silicate styles (Mica Group: "Biotite").

The iron is what gives the Biotite its dark color as compared to another mineral. Muscovite, which doesn't have iron in it, has a transparent appearance similar to cellophane.

**Orebites**: economically important ore minerals. Minerals such as iron, copper, gold, silver, zinc, lead, and calcium are extracted from base ores through a variety of processes.

**Carbonates:** form the group of minerals that have calcium carbonate ($CaCO_3$) in their chemical formulas. Carbonates are a very important group of minerals. These are often found in shells and fossils. Rocks that have calcium carbonate in them are called carbonated rocks. Examples of carbonates include **Coral reefs** (composed of calcium carbonate) and **Coquina** (a biomass of shells compressed to form a carbonate mineral).

**Mineral Chemistry and Analysis**

Minerals are described by their chemical composition, which, within specific limits, can widely vary. These variable compositions consist of **end members** within a specific group/family/class, and the variability is reflected by the percentage of elements within a specific mineral. Albite ($NaAlSiO_3$) and anorthite ($CaAl_2Si_2O_3$) are both types of feldspar, and their chemical formulas represent the end members of their particular group. Note that they share a common composition of aluminum (Al), silica (Si), and oxygen (O). They differ in that albite also contains Sodium (Na) while anorthite also contains calcium (Ca).

The minerals themselves are divided into families and then further subdivided into groups. Mineral classifications within each of the groups reflect both chemical composition and crystalline structure.

**Mineral Families**: Native elements (i.e. gold), Sulfides, Oxides, Halides, Carbonates, and Silicates.

Minerals are crystalline solids. The structure reflects the repetitive, periodic array of atoms. This array- the **lattice**- forms the basis for the **unit cell**: the basic repeating unit in a crystal that possesses the symmetry and properties of the mineral. The angles present at the bonding points of the atoms determine the fundamental geometric shape of the crystal. These shapes are classified as either isometric (cubic), tetragonal, orthorhombic, hexagonal, monoclinic, triclinic, or rhobohedral.

The chemical analysis of minerals is often displayed in a variety of graphic formats that address chemical composition and mineral stability. The primary formats used are either a Chemical Composition diagram or a **Pressure-Temperature (PT) diagram**.

A chemical composition diagram is used to show the molecular percentage of elements that combine to form the multiple minerals within a particular mineral group system. The format of the diagram consists of a triangle with the native elements shown at the apex and lower corners.

Percentile marks are shown along the sides of the triangle and a series of dots and lines section off specific areas that reflect the percentage of native element present for each member of the group. The percentages present are read at the intersection point on the line.

A pressure-temperature diagram is often used to display the stability of a particular mineral or group of minerals. A PT diagram uses a basic graph format with pressure shown on the y-axis and temperature shown on the x-axis. It is read as you would read any standard graph.

**SKILL 19.2**         **Identify the processes by which different kinds of rocks are formed.**

### Sedimentary Rocks

Sediments are basically various sized fragments of broken or eroded rock material. As weathering processes break down the parent material, the sediments are transported, sorted and deposited in piles. These sediment piles may lithify into sedimentary rocks.

### Fragmentation, Transportation and Sorting

### Wind and Running Water

Wind is very effective in transporting sediments in areas of little vegetation. The haze over the Grand Canyon is actually sand from Monument Valley over 50 miles away.

Sand dunes are essentially piles of wind-blown micro-rock. Pebbles in mountain pools and streambeds are pieces of the mountain broken off and carried to another location by water movement. The pebbles are polished and rounded by the abrasive action of tumbling into each other as the water carries them downstream.

The piles of sorted and transported sediment can be quite extensive in terms of depth and area covered. Everything east of Richmond, Virginia is a sedimentary base. The sediments came from the erosion of the Appalachian Mountains.

The **velocity** achieved by most transport agents determines what and how far materials are transported, and those materials tend to be of relatively uniform size. The exception is those materials moved by glaciers. A **glacier** acts like a giant bulldozer, moving most everything in its path, big or small.

<u>Fining</u>: the process of sorting materials by size. Geologists describe sedimentary rocks by size. To determine the size, they use the **Wentworth Scale.**

**Wentworth Scale**

| Name | Size | Size Analogy |
| --- | --- | --- |
| **Boulder** | Larger than 256mm | Basketball |
| **Cobbles** | > 64mm and < 64mm | Tennis ball |
| **Pebbles** | > 2mm and < 64mm | Pea sized |
| **Sand** | 1/16mm to 2mm | Coarse/ med/ fine grains |
| **Silt** | 1/256mm to 1/16mm | none |
| **Clay** | Less than 1/256mm | none |

There may be a variety of various sized sediments present in any given area. This size diversity demonstrates that over time a variety of transportation agents laid down the sedimentary material.

Geologists can measure the size of smaller materials by passing the material through screens with different sized openings. The point of determining size is to be able to recreate past environments. Size can provide clues to the original location of the material. Sedimentation is a continuous process.

**Murky water:** water carrying silt and clay sediments (suspended load) from upstream. Further upstream the materials are coarser.

The flowing water carries the material to the ocean where one of two things happen: the material is deposited on the offshore continental shelf, or the material is carried back inland to the inlets and bays.

Over time, the sediment thickly accumulates and may form typical coastal features such as sand bars and deltas. The continuous accumulation of sediment is why there is a continuing need to dredge harbors and rivers.

The lithification processes of cementation, compaction/dessification, and precipitation form sedimentary rocks.

**Cementation:** Sedimentary materials deposited in a pile are of different sizes. Consequently they have different sized spaces between grains. Some of these spaces are large enough to permit water to flow through them (sand sized or larger have large spaces between them).

Groundwater has lots of chemicals in it. As water moves through the spaces, the pH – measure of acidity – changes, and, drop by drop, chemicals are **precipitated** (deposited) along the edges of the grains. The spaces eventually become filled, and the precipitated chemicals hold the materials together.

The two most common cementing agents are Silica and Calcium Carbonate. **Silica ($SiO_4$)** is very hard (7 on the Mohs scale). Silica forms a rock that is very hard to break. **Calcium Carbonate ($CaCO_3$)** is less resistant to weathering (the calcium carbonate is easily dissolved).

Chemical impurities in flowing groundwater may collect within the spaces between sediment grains. These impurities cement the grains together to form a rock.

**Compaction and Dessification**: the processes that affect fine grain sediments.

Silt, clay and extremely fine sand become compressed by the weight (pressure) of the dirt and other materials on top of them and eventually dry out.

**Precipitation:** Where chemicals in the groundwater precipitate as solutions in the underground cracks and crevices. Precipitation is the result of chemicals, originally deposited by groundwater, that have dried out on the surface of the material. Example: Water spots that harden on the shower walls. The two most common precipitatory agents are Silica and Calcium Carbonate. Calcium Carbonate formations are often found in caves, in the form of Stalactites and Stalagmites. **Stalactites** form from the roof. **Stalagmites** form on the floor. (Memory jogger: the "g" means ground up!)

The **Geode** is an example of silica precipitation. Quartz crystals form inside of a rock exterior. Precipitation doesn't occur between sediment grains, instead, whole cavities in existing rock material are filled.

## Igneous Rocks

Igneous rocks form from the cooling and crystallization of a rock melt. **Melt:** the overall collective term used to describe molten and semi-molten rock material in the Earth. The melt is due to the heat present in the Earth and this heat is derived from two sources: the decay of radioactive elements and frictional forces and pressure within the Earth. As you go deeper into the Earth, the temperature increases. The various heat layers are collectively referred to as the **Geothermal Gradient**. Other forces can aid or hinder the melt.

**Pressure:** Rock stays solid for longer time periods if it is under pressure.

**Presence of Water:** Dependent on the pressure, the presence of water can delay or accelerate the melt process.

As rock materials move within the Earth, rocks in a liquid state move upward, seeking cracks in solid rock, and rocks in a solid state move downward.

Liquid rock slowly cools as it moves upwards. The upward movement can cause enough pressure and stress to move and/or fracture the solid rock.

## Crystallization

Magma is rich in chemicals. As it cools, chemicals combine to form distinct mineral structures. As minerals form they either settle to the bottom of the magma well or continue to react and form a richer magma.

**Bowen's Reaction Series** is used to predict how and which minerals will solidify out of a melt. These predictions are primarily based on the heat of the magma.

Lavas on the surface have a temperature of around 2000 °F. By comparison, magma that cools underground can take millions of years because of the extremely high temperatures present below the Earth's surface.

**Slow cooling:** forms very large crystals of minerals in the rock. Very large grains are prominent. Example: Granite.

**Quick Cooling:** This allows less time for the crystal to cool and form. It results in smaller (sometimes microscopic) crystals. Example: Basalt.

**Squelching (Instant Cooling):** This cooling occurs almost instantaneously when lava flows into the ocean or is thrown into the air by an eruption. No crystals are formed and the rocks usually have a glassy appearance. Example: Obsidian.

The cooling rate of the igneous rock is very important because it produces a distinct texture that is a key factor in its identification.

## Textural Results

**Phaneritic:** Igneous rock with mineral grains large enough to be seen with the naked eye. Grain size caused by slow cooling.

**Aphanitic:** Igneous rock with mineral crystals present, but the crystals are often too small to be seen without the aid of a microscope. Grain size caused by quick cooling.

## Categories of Igneous Rocks

**Extrusive:** Igneous rocks that cool on or near the Earth's surface. Characterized by Aphanitic or Glassy textures.

**Intrusive:** Igneous rocks that cool deep within the Earth. Characterized by Phanaretic texture. Intrusive rocks never reach the surface during the cooling process. Instead, they are exposed after millions of years of weathering. Examples include Stone Mountain in Georgia or Half Dome in Yosemite National Park.

**Glassy Texture:** no detectable mineral crystals present. Caused by squelching (instant cooling).

**Porphyritic:** some large mineral crystals within an Aphanitic or Glassy ground mass (background). The background is also known as the **matrix** and presents a Checker-board texture. The magma cooled slowly, forming large grains, but before the magma was fully developed, it erupted upward through cracks, cooled quickly, and other minerals formed smaller crystals as the matrix. Example: Diamond Head in Hawaii is composed of volcanic basalt with large Olivine crystals.

## Metamorphic Rocks

**Metamorphism:** changing a pre-existing rock into a new rock by heat and or pressure. Metamorphism is a process that is similar to that of putting a clay pot into a kiln. The clay doesn't melt, but a solid-state chemical reaction occurs that causes a change. The chemical bonds of adjoining atoms breakdown and allow the atoms to rearrange themselves, producing a substance with new properties.

**Single Mineral Metamorphism:** If the pre-existing rock is composed primarily of only one mineral, the metamorphic result is a rock with the same composition, but the crystal grains are larger and interlocked. For example, Sandstone is a cementation of silica. Under metamorphosis, the grains become larger and are fused together to form Quartzite. There are major differences in appearance and properties. Quartzite appears crystalline with very large crystals and is very hard. Another example: Metamorphism causes Limestone to become Marble. Limestone can be cut and used as building material, but it resists polishing, restricting its use as a decorative stone. Marble can be polished and utilized as very strong and decorative building material.

**Multiple Mineral Metamorphism:** If the pre-existing rock is composed of more than one mineral, then pre-existing minerals may align to give a new appearance, or recombination can occur within the rock, producing entirely new minerals. An example of this is the metamorphic change of Granite. It retains the same mineral composition but becomes Gneiss, ending up with the materials aligned, and giving it a striped appearance. Likewise, Garnet is a recombination of multiple minerals and is formed only by metamorphism.

Temperature and/or pressure cause metamorphism. The metamorphic effect may produce a change in the chemical and physical properties and/or appearance of the rock.

**Types of Metamorphism**

**Contact Metamorphosis – Temperature:** This requires the presence of a nearby magma chamber. The closer to the heat source, the more metamorphosis that takes place. This is a localized effect due to the presence of a magma chamber. The effect is measured in 10's to 100's of yards. Rocks formed during contact metamorphism tend not to be foliated (striped).

**Regional Metamorphism – Deformation through pressure:** This type of metamorphism is produced by the tectonic movement (drift) of continental plates. Continental drift exerts enormous pressure at the edges of the plates (where they abut). Rocks are forced into a semi-plastic state and the atoms break their bonds. The pressures involved cause the rocks to reform without breaking. An analogy for this type of metamorphism is that of chocolate chip cookies fresh and hot from the oven. Under pressure, they tend to bend rather than break. In the case of rock material, pressure compression causes the material to fold. Regional metamorphism has a widespread effect measured in 10's to 100's of miles. Example: The Appalachian Mountains have abundant metamorphic material because the mountains were the result of three collisions in the past between the North American and African plates. In regional metamorphism you end up with deformed rocks that are highly foliated.

**Migrating Fluids – (Also called Metasomatism or Hydrothermal Alteration):** Formed by groundwater heated by a magma chamber. Heated water (near steam) moves through cracks in the rocks. Over time, the water leeches minerals out of the rock. These minerals are concentrated in solution in the water and are transported to other locations to precipitate as a vein. The effect of migrating fluids tends to be localized and is measured in 10's to 100's of yards. This type of metamorphism is usually associated with the presence of large plutons.

**Index Minerals:** select minerals used to measure the metamorphic grade, which reflects the intensity of the metamorphic process. The presence of a particular index mineral provides information about the temperature and pressure involved in a metamorphic event. As a general rule of thumb: the closer to the heat source, the greater the metamorphic effect.

**SKILL 19.3    Classify a given rock as sedimentary, igneous, or metamorphic.**

<u>Rock</u>: two or more minerals combined to form a single entity. There are three main categories of rocks: Igneous, Metamorphic and Sedimentary.

**Igneous Rocks:** rocks that are formed from a melt, which is created by extreme pressure and temperature deep below the surface of the Earth.

Material in a melt is referred to as either **Magma**, melted rock material below ground, or **Lava**, melted rock material above ground.

As lava cools and solidifies, minerals arise in combinations based upon their respective densities.

Igneous rocks are classified by texture, size of grain, and chemical composition.

**Igneous Intrusion:** melted rock squeezed (intruding) into cavities in existing rocks. These intrusions form **Plutons**: bodies of igneous material deep within the Earth. **Batholith:** a very large pluton. Batholiths are usually solidified magma chambers that can be up to 100 miles or more across. Example: Stone Mountain in Georgia.

**Dike:** a small, linear shaped pluton. Pressure forces magma upward through fault cracks, enlarging the cracks. The dikes can also go sideways into bedrock cracks. Example: Ship Rock in Arizona.

**Lacolith:** a dome or mushroom shaped pluton, much smaller than a Batholith.

**Metamorphic Rocks:** rocks that undergo a Metamorphisis (change). The change is caused by heat and/or pressure. The degree of metamorphism depends on the magnitude of the heat and/or pressure applied. In this process, the rock doesn't melt; it is baked into a new substance.

**Classification of Metamorphic Rocks:**

Metamorphic rocks are classified by texture (appearance). They are either Foliated or Non-Foliated. **Foliated**: a banded (striped) appearance to the naked eye. Foliation usually results from regional metamorphism. This classification is further sub-classified by the degree of foliation. Some examples include:

> **Gneiss:** Very obvious striped banding.
>
> **Schist:** "Flaky" appearance. Tends to be shiny like glitter stuck all over the rock or the sheen from fish scales.
>
> **Slate:** Cleavage with a slight sheen. Slate looks like shale but does not have a "muddy" appearance, and the edges appear "peelable" at the cleavage. Harder than shale, slate "clinks" when struck. Shale "clunks" when struck.

**Non-Foliated (also called Massive):** no foliation. Examples: Quartzite or Marble.

**Metamorphosis: Coal to Diamond**

Although actually classified as a sedimentary rock because it is composed of organic materials, coal is considered to be a special type of metamorphic rock because it is formed by repeated metamorphosis. Coal starts as organic material - such as leaves- falling to the ground. Eventually, the organic material is buried to the extent that it doesn't decay because of the oxygen poor environment. Large accumulations of non-decayed organic matter usually form sedimentary rocks due to compression/dessification processes. However, continued pressure on the organic accumulation can cause a metamorphosis, turning the organic matter into lignite. Lignite is a type of pseudo-coal. It burns, but it is very smoky and gives off low heat. Under greater pressure, lignite turns into Bituminous Coal. This is a soft coal that produces low temperatures when burned. When placed under more pressure, bituminous coal changes into Anthracite Coal. This is a very hard coal that produces high heat when burned. Under still more pressure, metamorphism changes anthracite coal into Graphite. Finally, under still higher pressure, graphite turns into Diamond.

**Sedimentary Rocks:** rocks that form from sediments that lithify to become solid rock. **Sediments** are broken up rock material. Sand on a beach or pebbles in a mountain stream are typical examples.

Lithify: to become a rock. A classic example of this process is sandstone. When sand particles are placed under great pressure, they lithify to become sandstone.

**Petrology**: the study of how rocks are formed.

**Classification of Sedimentary Rocks**

Classification of sedimentary rocks is based upon the way the rock was formed.

  **Castic (Fragments)**: Composed of lithified sediments (fragmental), which have undergone Compaction/Dessification or Cementation.

  **Conglomerate**: Large, bigger than sand, rounded particles. Looks like a pile of pebbles stuck together.

  **Breccia**: Large, bigger than sand, angular particles stuck together. Have sharp edges and sides.

  **Sandstone**: Sand sized particles. The feel and size is dependent on the size of the sand dune that it came from.

  **Shale**: Less than sand sized particles. Clay and silt sized. Grains are very fine, can't be seen with naked eye.

Conglomerate, Breccia and Sandstone are formed by cementation. Shale is formed by compaction/dessification.

**Bioclastic or Organic**: Formed from the remains of once living organisms that were usually found in marine environments. The temperature of the water dictates the type of bioclastic collections. Cold = Deepwater = Silica. Warmer = Shallow water = Calcium Carbonate.

**Coquina** is composed of large shell fragments cemented together. Very easy to identify, Coquina was a prized building material in the early settler days. Example: Old Spanish forts in Fort Lauderdale, Florida.

**Fossiliferous Limestone:** Composed of calcium carbonate or carbonate materials, fossils are evident, often it is reef material.

**Chalk:** Composed of microscopic calcium carbonate shells. Accumulations can be quite thick. Example: The White Cliffs of Dover in England are chalk deposits 2000 ft in depth.

**Chert:** Microscopic silica shells.

**Non-Clastic or Chemical:** Precipitated out of a solution. This is a very slow process, as the rock is formed drop-by-drop.

**Geode:** Silica collected inside a cavity in other rocks.

**Ootitic Limestone:** Calcium carbonate material that precipitates directly from seawater.

**Agate:** Banded silicate materials with light and dark colors.

Often, a specific mineral or mineral series is associated with a particular type of rock.

### Common Rock-Mineral Associations

| Rock | Mineral |
|---|---|
| Schist | Kyanite, Tourmaline |
| Gneiss | Kyanite, Quartz |
| Granites | Beryl, Tourmaline, Biotite, Feldspar (potassium & plagioclase), Quartz |

However, the formation of the igneous rock is dependent on a variety of conditions (including location and presence or absence of water), with pressure and temperature playing key roles in determining what type of igneous rock is formed.

Granite and Granodiorites are the most common intrusive igneous rock in the continental crust. Andesite is the dominant extrusive rock associated with subduction zone volcanism. Basalt is the most abundant extrusive rock, and it is associated with oceanic crust, rift zones, and oceanic and continental hot spot volcanism. Peridotite dominates the Earth's upper mantle and most magma that intrudes the crust at rift zones, rift valleys, and oceanic ridges. Feldspar remains the most abundant minerals in the silicate rocks.

**COMPETENCY 20.0** Understand the structure of the Earth, the constructional forces that have shaped its surface, theories and evidence of crustal movements, and the effects of crustal movements on landscape.

**SKILL 20.1** Use seismic waves to infer the Earth's internal structure.

**Seismology**: the study of earthquakes.

Seismic waves are measured using a device called a Seismometer, a type of motion sensor. The seismometer is anchored to the Earth and a heavy weight is suspended on its frame.

As the Earth materials move, the weight also moves and electronically sends a signal to a recording device called a Seismograph. Movements are displayed as a series of lines on a recording chart called a Seismogram, reflecting the seismic energy detected at a particular location.

**Types of Seismic Waves**

P-Wave (Primary Wave): Also sometimes called a "Push-Pull Wave or Compression Wave," the P-Wave moves through both solids and liquids. The P-Wave has a pulsating, "push-pull" type motion. It compresses material as it moves through it. The fastest moving of the seismic waves (4-7 Km/sec), the P-Wave is the first wave to reach the seismometers.

S-Wave (Secondary Wave, also called a Shake Wave): Moves only through solid material. Always shorter than a P-Wave, the S-Wave (2-5 Km/sec) is the second wave that reaches the seismometer. Motion is a sinuous "side to side" movement.

L-Wave (Surface Wave): The L-Wave is much slower than either the P- or S-Waves, but creates lots of ground movement. Because it is slower, the L-Wave takes longer to pass a location and consequently, the intense, undulating ground motion creates the greatest amount of damage in an earthquake. The L-Wave undulates with a rolling motion of the Earth, similar to ocean waves.

**Measuring Magnitude and Intensity**

**Magnitude**: the relative measure of how big an earthquake is- how much energy is released.

**Intensity**: the measure of observable effect in terms of damage and destruction caused by an earthquake.

**Richter Scale**: the primary scale used by seismologists to measure the magnitude of the energy released in an earthquake. It is a logarithmic scale. *Example:* A 5.0 earthquake is 10 times stronger than a 4.0 earthquake, and 100 times stronger than a 3.0 earthquake. Although the Richter Scale is still the most widely known scale, it is only one of several types of scales used to measure magnitude. Example: The Fuji scale is gaining popularity. It is a more accurate measurement of the energy released by an earthquake.

Finding out the effects of an earthquake is useful in estimating the intensity of the earthquake. These effects include the extent of horizontal and lateral movement and the degree of deformation and destruction of the landscape and buildings.

The Modified Mercalli Intensity Scale of 1931 provides a correlation of observable damage to approximate intensity. A corresponding number indicates the degree of observed effects. The higher the number, the greater the intensity. A series of seismometers are used to locate the epicenter of an earthquake through a geometrical process called triangulation. A minimum of three seismometers is required to accurately triangulate the epicenter. Seismometers are also used to determine the magnitude and distance by plotting a travel-time curve, derived by measuring the time lag between to arrival of the P and S waves.

**SKILL 20.2** **Use the theory of dynamic equilibrium (e.g., isostasy) to explain landscape development.**

Landscape development demonstrates a repeating pattern of periods of uplift that provide energy for subsequent periods of denudation and plantation. Denudation is the process by which elevation and relief in landscapes is reduced through the removal of material by erosion and weathering. There are three main sources of uplift. These are orogenic uplift, epeirogenic uplift, and isostatic uplift.

Isostasy is the state of gravitational equilibrium between the Earth's lithosphere and asthenosphere that allows the solid tectonic plates to float on the more liquid asthenosphere at an elevation dependent on their thickness and density. Isostatic uplift is the rise in the level of land relative to the sea that is caused by the massive removal of weight from a particular area by erosion or ice-sheet retreat.

Large amounts of sediment deposition at a particular location may cause the Earth's crust to sink lower into the asthenosphere. Similarly, large amounts of erosion may cause a particular area to rise. These two processes generate a cycle in which mountain ranges and elevated areas erode down, remaining land rebounds upwards to create new elevations due to isostasy, and new elevations then begin to erode. It is through this cycle that landscape undergoes a pattern of growth and decline. Additionally, erosion and land rebound eventually expose rock strata from deep in the Earth to the surface, changing a landscape's shape and composition.

The formation of ice-sheets can also cause areas of the Earth's surface to sink. Isostatic uplift may also be referred to as post-glacial rebound, which is defined as the rise of landmasses that were depressed by the huge weight of ice sheets during the last ice age. Uplift in these areas occurs as temperatures rise, removing the weight of ice-sheets through melting and allowing landmasses to rise significantly. Post-glacial rebound affects northern Europe, especially Scotland and Scandinavia, Siberia, Canada, and the Great Lakes of Canada and the United States. The dramatic effects of post-glacial rebound may be seen in regions formerly covered by ice sheets, such as the Baltic Sea and the Hudson bay, where areas once constituting sea cliffs and beach ridges can now be found at 300 meters above sea level.

**SKILL 20.3  Describe the evidence for continental drift and seafloor spreading.**

**Support for Tectonic Theory**

**Shape of the Continents**: When graphically displayed, the continents look like they should largely fit together in a jigsaw puzzle fashion.

**Paleomagnetism**: As igneous rock cools, iron minerals within the rock will align much like a compass to the magnetic pole. Scientific research has shown that the magnetic pole periodically—hundreds of thousands of years—reverses polarity.
- Normal Polarity: Magnetic North.
- Reverse Polarity: Magnetic South.

Research also shows that the bands of rocks on either side of a spreading center are mirror images of each other with regards to magnetic polarity, and that the alignment of minerals indicates a periodic shift in polarity. The reversals in polarity can be visualized as alternating "stripes" of magnetic oceanic materials.

**Age of the Rock**: Besides being mirror images magnetically, dating research conducted on rocks on either side of a spreading center also indicate a mirroring of age. The age of the rock on either side of a spreading center are mirror images and get progressively older as you move away from the center. The youngest rock is always found directly at the spreading center. In comparison to continental rock materials, the youngest rock is found on the ocean floor, consistent with the tectonic theory of cyclic spreading and subduction. Overall, oceanic material is roughly 200 million years old, while most continental material is significantly older, with age measured in billions of years.

**Climatology**: This is one of the most compelling arguments supporting plate movement. Cold areas show evidence of once having been hot and vice-versa. In example, coal needs a hot and humid climate to form. It does not form in areas of extreme cold. Although Antarctica is extremely cold, it has huge coal deposits. This indicates that at one time in the past, Antarctica must have been much closer to the equator.

**Evidence of Identical Rock Units**: Rock units can be traced across ocean basins. Many rocks are distinctive in feature, composition, etc. Identical rock units have been found on multiple continents, usually along the edges of where the plates once (apparently) joined. Example: A significant number of South American and African rock unit formations are identical.

**Topographic Evidence**: Topographic features can be traced across ocean basins. Some glacial deposits, stream channels, and mountain ranges terminate on one continent near the waters' edge, and resume on another continent in relatively the same position.

**Fossil Evidence**: Limited range fossils that could not swim or fly are found on either side of an ocean basin. Example: The fossilized remains of pandas, kangaroos and the long extinct Metasaurus are unique to only two areas of the world. Ocean basins separate these areas.

**Sea Turtle Migration**: The genetic instincts of Sea Turtles drive them to return to the islands where they hatched in order to lay eggs. Migration of the sea turtles over thousands of miles is well documented. The diverse location and number of islands to which the sea turtles migrate suggests that the plate movement has changed the location of the islands from their original position immediately off shore of major continental masses.

**SKILL 20.4** **Apply the theory of plate tectonics to explain landscape development and geologic phenomena (e.g., volcanism, earthquakes) and to predict future movements of land masses.**

**Tectonic Plate Movement**

Plate tectonic movement results from the motion induced in the lithosphere by the rise and fall of convection cell material in the asthenosphere.

**Plate Boundaries**: the points at which the edges of tectonic plates abut.

Three motions characterize interactions at the plate boundaries: separation, collision, or lateral movement. Those motions directly correlate with the categorizations of plate boundaries; Divergent, Convergent, and Transform. The geologic and geographic effects that result from the motion depend on the location of the boundaries and the types of material involved.

**Divergent Boundary**: The plates are separating and moving away from each other.

Ocean/Ocean Boundaries: The materials involved are composed of heavy and dense, but very thin, dark colored oceanic lithospheric material, usually Basalt. As the magma rises, the ocean floor begins to dome upward. The upward pressure eventually forces an underwater rip in the center of the dome and the magma erupts. The erupted materials cool rapidly and build upward, forming Mid-Ocean Ridges, which are fairly common and found all over the globe. The ocean floor is constantly being pushed apart at these boundaries, causing Sea Floor Spreading. This results in the creation of huge oceanic plates.

Continental/Continental Boundaries: The materials involved are composed of less dense, but very thick, lighter colored continental lithospheric material, usually Granite. The same principles of force and motion present in divergent ocean/ocean boundaries apply to divergent continental/continental boundaries. However, as the erupted material eventually cools, a Rift Valley forms between the adjoining volcanic peaks. In many places around the world, these valleys play an important agricultural role because of their rich volcanic soil.

**Convergent Boundaries**: The plates are moving toward each other and collide.

Ocean/Ocean Boundaries: These plates are forced together by the spreading of the ocean floor. Tremendous frictional forces are created as the plates collide and some of the oceanic material builds upward, while other oceanic material bends downward. The leading edges of the boundaries meet around 700 km downward. The forces involved push some material upward through the lithosphere to become a volcano. The built up materials may eventually break the ocean surface to become volcanic islands. This effect is so widespread that the islands form groupings of volcanic islands called Volcanic Arcs. This volcanism is of the explosive type and the quick release of fantastic strain and pressure causes devastating Deep Focus Earthquakes. Although some material is pushed upward, other oceanic material bends downward forming deep trenches and the leading edges of this plate will subduct back into the asthenosphere.

Subduction Zone: A long, narrow belt where a lithospheric plate dives into the asthenosphere. The rate of subduction is relatively equal to the rate of formation of new oceanic lithospheric material at divergent boundary spreading centers. In effect, the ocean floor recycles itself.

Ocean/Continental Boundaries: The colliding plates produce effects relatively similar to ocean/ocean collisions, but the difference in density between the materials involved causes the oceanic plate to subduct under the continental plate. Subduction forces the continental materials upward, creating a line of onshore volcanic mountains along the subduction zone.

Continental/Continental Boundaries: Both edges are too light to subduct. Instead, one will over ride the other, causing an uplift of material.

**Transform Boundary**: The plates move laterally with respect to each other. As the plates grind sideways, intense frictional forces are created as the lithospheric materials try to oppose the movement. A transform boundary may be found in any location where plates abut. They may be composed of any type of lithospheric material (oceanic or continental), and they produce extreme seismic effects when the pressure between moving boundaries is released. This sudden release of pressure creates widespread destruction along the fault lines.

**Hot Spot/Aseismic Ridge**

Hot Spot: an offshoot of a convection cell that burns through the lithospheric material in the middle, not at the edges, of a plate.

Aseismic Ridge: a series of volcanoes (underwater or above the surface) formed by the movement of the plates over a hot spot. They usually form a dogleg pattern, showing the change of direction in the plates' movement.

**COMPETENCY 21.0**     **Understand erosional-depositional processes that change the Earth's surface (e.g., weathering, erosion) and the relationship between these processes and landscape development.**

**SKILL 21.1**     **Describe the processes of mechanical, chemical, and biological weather and factors that affect the rate at which rocks weather and soils are produced.**

**Weathering**: the physical and chemical breakdown and alteration of rocks and minerals at or near the Earth's surface.

**Mechanical (also called Physical Weathering):** where rock is broken into smaller pieces with no change in chemical or mineralogical composition. The resulting material still resembles the original material. Example: Rock pieces breaking off of a boulder. The pieces still resemble the original material, but on a smaller scale.

**Chemical Weathering:** where a chemical or mineralogical change occurs in the rock and the resulting material no longer resembles the original material. Example: Granite (Gneiss/Schist) eventually weathers into separate sand, silt, and clay particles.

Weathering is typically caused by a combination of chemical and mechanical processes.

**Factors Influencing Weathering**

Composition: Due to their composition, some rocks weather easier and will show more effects.

Rock Structure: Does it have cracks? Is it fractured? Water and other elements get into the cracks.

Climate: The more water, the more the weathering effect. Additionally, the higher the temperature or the more the temperature varies, the greater the weathering effect.

Topography: This factor determines the amount of surface area exposed to weathering. Smaller rocks are affected more because, collectively, they have less mass and more surface area than a boulder.

Vegetation: Important weathering agent. Depending on the type of vegetation, it can either hinder or accelerate the weathering process. Although vegetation may leave less surface area exposed, the vegetation's root structures can produce a biological effect that accelerates the process.

**Types of Mechanical Weathering**

Frost Wedging: This occurs when a rock has a crack in which water collects and then freezes. Over time, as this cycle repeats itself, the expanding water gradually pushes the rock apart.

Salt Crystal Growth: In a process similar to frost wedging, as the water evaporates, it leaves salt crystals behind. Eventually, these crystals build up and push the rock apart. This is a very small-scale effect and takes considerably longer than frost wedging to affect the rock material.

Abrasion: This is a key factor in mechanical weathering. The motion of the landscape materials produces significant weathering effects thereby scouring, chipping, or wearing away pieces of material. Abrasive agents include wind blown sand, water movement, and landslides.

Biological Activity: This is a two-fold weathering agent.

> Plants: Seeds will sometimes land in a crack in a rock and begin to grow in the crack. The root structure eventually acts as a wedge, pushing the rock apart.

> Animal: As animals burrow, the displaced material has an abrasive effect on the surrounding rock. Because of the limited number of burrowing animals, plant activity has a much greater weathering effect.

Pressure Release (Exfoliation): Rock expands when compressive forces are removed, and bits of the rock break off during expansion. This can result in massive rock formations with rounded edges. Example: Half-Dome in Yosemite National Park.

Thermal Expansion and Contraction: Minerals within a rock will expand or contract due to changes in temperature. Dependent on the minerals in the rock, this expansion and contraction occurs at different rates and to different magnitudes. Essentially, the rock internally tears itself apart. The rock may look solid but when placed under pressure it easily crumbles.

## Types of Chemical Weathering

Oxidation (Rust): Oxygen atoms become incorporated into the formula of a mineral in a rock and the mineral becomes unstable and breaks off in flakes. Example: Iron oxide ($FeO_2$) changes to iron trioxide ($FeO_3$) due to the oxygen chemically imparted to the mineral.

Solution:  Due to their inherent composition, some minerals found in rocks easily dissolve into solution when exposed to a liquid.
Example:  Halite (Rock Salt) completely dissolves in water.

Acids:  Water and water vapor may combine with other elements and gases to form acids.  Water ($H_2O$) and carbon dioxide ($CO_2$) can chemically combine to become Carbonic Acid ($H_2CO_3$).  Sulfur Dioxide ($SO_2$) particles can chemically combine with water ($H_2O$) to form Sulfuric Acid ($H_2SO_4$). Generally found in combination with solution, acids cause the majority of chemical weathering.

Biological Activity: Plant roots growing in the cracks of rocks not only cause mechanical wedging, but also secrete acids that cause chemical weathering.

## Differential Weathering

Differential weathering occurs when rocks on a landscape weather at different rates.  The effect of this difference can create fantastic landscape features, with some rocks sticking up, arched, or nested in deep depressions.  Examples include Ship Rock, Devil's Tower, and the other mesa and butte formations in Monument Valley.

## Sediments and Soils

**Sediment:**  fragments of broken rock produced by the weathering process.

**Soils** form when sediments undergo the process of leeching, accumulation, and addition of organic matter.  Soil has layers (horizons) of mineral material and organic matter that differ substantially from the parent material.

**Parent Material:** the rock or sediment from which a soil is derived.

## Composition of Soil

Soil is comprised of layers called horizons.  Each horizon represents a particular stage of soil formation and the presence and depth of each horizon is unique to the area that the soil is found.

**O Horizon:** This layer is comprised of mostly litter and humus and has a very small proportion of minerals. The O horizon is named for its organic content and is the uppermost layer of the soil profile.

**A Horizon:** A mixture of humus, sand, silt, and clay, this layer has a much higher percentage of minerals.

The O and A horizons comprise the **Topsoil:** the fertile layers of the soil profile that contain millions of living organisms.

As rainwater travels through the topsoil, it partially dissolves the minerals and salts and carries them downward to the "B" horizon. This process is called leaching, and thus, the "A" horizon is known as the **Zone of Leaching.**

**B Horizon: The subsoil.** It is a transitional zone of partially weathered bedrock and very little organic material. This horizon is known as the **Zone of Accumulation** because this layer is the collection point in the soil for the leached minerals, sand, and clay from the "A" horizon.

**C Horizon:** This layer consists of the **parent material.**

There is no set depth for soil. The extent and depth of the individual horizons will vary widely according to geographic and geologic region.

**SKILL 21.2** **Identify the processes of erosion by various agents (e.g., wind, water, glaciers) and factors that affect erosion rates and patterns.**

Wind and water carve landscapes. Landscapes evolve as wind and streams transport land, cut away land, and deposit it elsewhere.

**Landscapes and Streams**

Streams progress and evolve over time. In land that has been uplifted, water cuts down through the rock in a straight path, forming what is known as a young stream. The characteristics of a young stream include a straight, narrow, V-shaped channel that is actively cutting down through rock layers. Sediment is transported and deposited. This process decreases the slope, which decreases the velocity of the water. As the velocity decreases, the stream begins to meander, forming a mature stream. Sinuosity is the measure of how straight or curved a stream is. It is the ratio of the actual channel length and the distance down the valley. A young stream has a sinuosity of 3 and under. A mature stream has a sinuosity of 4 and over.

Young streams change the landscape as they cut down through the rock, lessening the slope. Mature streams alter the landscape as they meander through the land. The outside of the curve of a meander is called a cut bank. The stream has the highest velocity there, and cuts away increasing the curve. The point bank is on the inside of the curve. The velocity of the stream slows on the inside of the curve, depositing the sediment. The position of the stream moves over time as land is deposited in the point bar, and the cut bank is actively eroded. As a meander becomes more U-shaped, there is a possibility for the stream to abandon the curve and follow the path of least resistance, forming an oxbow lake.

Floodplains form along mature streams as well. This process creates a wide, flat valley. As the banks overflow during a flooding event, large sediment materials are deposited at the banks, forming natural levees. Silt and clay materials are deposited on the floodplain, flattening the land.

Deltas and alluvial fans form when sediments are deposited as water slows quickly. An alluvial fan forms off of a mountain's slope into the valley as the velocity slows rapidly in the valley. A delta forms when sediments are deposited quickly as a stream slows when it hits the ocean or other large body of water.

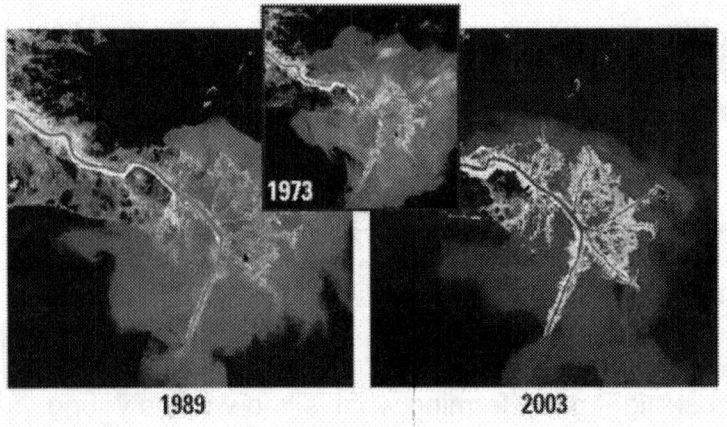

*http://eros.usgs.gov/kids/documentation/*

## Landscapes, Uplift, and Erosion

Through the process of plate tectonics some land is actively being uplifted (such as the Himalayan mountains). In these young mountains the relief is high and the peaks are very steep. The geomorphic evolution is the succession of young mountains, to mature mountains, to old age plains. This process is active. As mountains are forming, erosion of the soft layers is also taking place. The resistant layers remain. Once tectonics cease, weathering and erosion of the land will continue. The time needed for these mountains to mature depends on the rock type and the climate of the area. After a very long time, flat, low lying plains called peneplains form. Occasionally, resistant rock remains forming a single mountain or hill on a peneplain called a monadnock. The evolution of the land is through exposure to the elements: ice, water, wind, plant life, and temperature changes.

## Glacial Erosion

Glaciers are nature's bulldozers. As the glacier moves forward due to basal sliding, the underlying topography is severely abraded because of the immense pressure on the rock base due to the weight of the glacier.

The key factor in the glacial erosion process is the meltwater. It causes frost wedging, which initiates the erosional sequences.

As the glacier moves forward on the meltwater, some of the water seeps into cracks and freezes. This frost wedging causes a further widening and weakening of the rock material. The glacier plucks fragments out of the cracks and pushes them along the base. These fragments act as a scouring pad, and as they increase in size and amount, they increase the abrasion on the topography. The grinding motion due to the glacial weight increases as the mass moves forward, picking up more fragments and causing **striations** in the underlying rock.

Glaciers can move rock fragments varying in size from pebbles to entire boulders. Fine fragments are referred to as **rock flour**. Boulder sized rock units moved to other locations are called **eratics**.

SKILL 21.3    Explain the processes by which given landscape features (e.g., eskers, moraines) are formed.

## Erosional Features

### Alpine Valleys

After a glacier melts, it leaves behind its distinct mark on the landscape. One of the most common of these features is the **Alpine Glacier Valley**. These are usually easy to recognize. A characteristic of glacial landscaping is the U-shaped valley. These usually are situated in previously carved areas caused by streams.

However, the glacier's movement straightens out the curves normally associated with streams and also carves the sides of valley, neatly removing or dramatically shortening any ridges that extend in its path. **Truncated Spurs** – the lowest part of the intruding ridges - are the result of this movement.

Glaciers can also form tributaries similar to streams. Depending on the size of the glacier, the heavier ice masses can carve downward more rapidly and create much deeper valleys.

Once these glaciers recede, the tributaries remain behind, perched high above the main valley as **hanging valleys**.

Another typical feature of glaciation is the **rock basin lake**. These are water filled depressions carved in the bedrock by the abrasive action of a glacier's movement. These lakes fill after the glacier melts.

### Cirques, Horns and Aretes

**Cirque:** a steep-sided, rounded hollow at the head of a glacial valley. The cirque is one of the most prominent features associated with Alpine style glaciation.

**Horn**: the sharp peak formed by the erosional action of cirques. The mountain is cut back on several sides, forming the prominence.

**Arete**: a sharp ridge that separates the valleys carved by glaciers.

### Continental Erosional Features

**Till**: the fragments that are abraded from the bedrock and deposited along the path of the glacier. The till are usually angular shaped and unsorted.

**Moraine**: a body of till carried along or deposited by a glacier. This loose material falls from the side of the glacier and accumulates along the edges of the ice path in the valleys.

**Lateral Moraine**: piles of till along the sides of a glacier.

**Medial Moraine**: a single long ridge of till carried down the central path of a glacier.

**End Moraine**: a ridge of till that piles up at the edge of the ice sheet.

**Terminal Moraine**: a type of end moraine that marks the furthest advance of the glacier.

**Recessional Moraine**: a type of end moraine that marks the temporary terminus of a receding glacier.

**Ground Moraine**: an extensive, thin layer of deposited till that creates the rolling topography of areas once affected by glaciation.

**Drumlin**: a streamlined hill of till shaped like the bowl of an inverted spoon. The long axis of the drumlin is parallel to the direction of the ice movement.

**Meltwater Erosional Features**

A significant amount of meltwater runs over, beneath, and away from the ice sheet in the zone of wastage. The material deposited by this meltwater is known as **Outwash**.

The materials carried by the outwash behave similar to those moved by a stream; the material is sorted and layered, differentiating it from the unsorted till.

An **Esker**: a long, sinuous ridge comprised of outwash deposited sediment that can be up to 10 meters in height. The material is deposited in tunnels beneath the ice sheet and left behind as the ice melts.

As the outwash moves sediment alongside and in the path of a receding glacier, blocks of ice can be buried beneath the sediment. After years of erosion these blocks are uncovered and melt, leaving a shallow depression behind. When these depressions fill, they are known as **Kettles**, and become scenic lakes (Kettle Lakes).

The outwash carries a large amount of rock flour and when this eventually settles out and dries, it can be picked up and carried long distances by the wind. This fine-grained dust is called **Loess**.

Sometimes a lake forms between retreating glaciers and end moraines. Clay and silt settle on the bottom in two distinct layers. The light colored layer (silt) represents coarser sediment deposited by glacier melting during the warmer part of the year. The darker colored layer (clay) represents the more slowly settling material that sinks after the lake surface freezes. Taken together, the two-color pattern is called a **Varve**, and indicates the total deposition of material in a one-year span. Like tree-rings, varves can be used as a relative dating technique to roughly estimate the amount of time that the lake has existed.

**SKILL 21.4**      **Describe the effects of glaciation on the Massachusetts landscape.**

The Pleistocene epoch was characterized by four major glacial and inter-glacial cycles. During periods of lower temperatures, huge ice sheets advanced to cover all of Southern New England. During the last glacial period, known as the Wisconsinan glaciation, the Laurentide ice sheet reached its terminus approximately 25,000 years BP. The Laurentide ice sheet's terminal moraines generated the outer islands of Nantucket and Martha's Vineyard, marking the furthest extent of the Wisconsin glaciation in southeastern New England. By approximately 14,000 years BP, the glacier began to retreat, and had left New England by approximately 13,000 years ago.

Advancing and retreating glaciers of Massachusetts picked up the bedrock of Southern New England and deposited it as the glacial and postglacial sediments of Cape Cod. This rock debris, known as drift, was deposited along the ice front. As sea levels rose with warmer temperatures, drift along the shoreline was eroded and re-deposited to form the beaches and spits of the Massachusetts coastline. Dunes were formed as winds carried and re-deposited sand atop sea cliffs.

The glacier covering southeastern Massachusetts was formed into lobes (rounded projections) by basins in the underlying bedrock. These lobes occupied the present sites of Buzzard's Bay, Cape Cod Bay, and the Great South Channel to the east of Cape Cod. Re-advances of the Buzzard's Bay and Cape Cod Bay lobes against ice-contact margins of the Mashpee Pitted Plain and the Barnstable Outwash Plain formed Cape Cod's most prominent landforms, the Buzzard's Bay and Sandwich Moraines. These moraine surfaces are riddled with ridges, knobs, and kettles. The inner part of Cape Cod possesses forward sloping outwash plains. These plains were formed by streams that drained the Cape Cod Bay lobe. The outer part of the Cape has westward sloping outwash plains formed by meltwater that drained the South Channel lobe.

Another common glacial landform is the drumlin, an elongated whale-shaped hill that forms with an axis parallel with the movement of the ice. Drumlins usually consist of layered glacial till or rock, repeatedly added to over time. The Boston Harbor islands are coastal extensions of Massachusetts's asymmetrical, elongate drumlins that were formed into smooth-sloped hills on the Boston Basin lowlands.

## COMPETENCY 22.0 Understand characteristics of the major geologic time divisions and theories and supporting evidence of the Earth's geologic history.

## SKILL 22.1 Identify the conditions and characteristic fossils of the various geologic periods.

With over 600 genera and 1000's of species, Trilobites were bottom dwellers (scavengers) found in shallow to deep water. For an extremely long period of time, Trilobites were the dominant multi-cellular life form on the planet. Trilobites are very good guide fossils because they were extremely abundant and existed throughout the entire Paleozoic period. Their development underwent distinctive changes, and these differences are useful in subdividing the time period.

**Geologic time for emergence of life**

Archean: Early bacteria and algae
Cambrian: Early shelled organisms
Cambrian/Ordovician: Early fish
Silurian: Early land plants
Devonian: Early amphibians
Devonian/Carboniferous: Early trees
Carboniferous: Early reptiles
Late Triassic: Dinosaurs
Jurassic: Early birds and mammals
Cretaceous: early flowering plants
Cretaceous - Cenozoic boundary (KT): Extinction of the dinosaurs
Paleocene: Early primates
Eocene: Early horses
Pleistocene: Neanderthal man
Holocene: Modern man

| EON | ERA | PERIOD | | EPOCH | | Ma |
|---|---|---|---|---|---|---|
| Phanerozoic | Cenozoic | Quaternary | | Holocene | | 0.01 |
| | | | | Pleistocene | Late | 0.8 |
| | | | | | Early | 1.8 |
| | | Tertiary | Neogene | Pliocene | Late | 3.6 |
| | | | | | Early | 5.3 |
| | | | | Miocene | Late | 11.2 |
| | | | | | Middle | 16.4 |
| | | | | | Early | 33.7 |
| | | | | Oligocene | Late | 28.5 |
| | | | | | Early | 33.7 |
| | | | Paleogene | Eocene | Late | 41.3 |
| | | | | | Middle | 49.0 |
| | | | | | Early | 54.8 |
| | | | | Paleocene | Late | 61.0 |
| | | | | | Early | 65.0 |
| | Mesozoic | Cretaceous | | Late | | 99.0 |
| | | | | Early | | 144 |
| | | Jurassic | | Late | | 159 |
| | | | | Middle | | 180 |
| | | | | Early | | 206 |
| | | Triassic | | Late | | 227 |
| | | | | Middle | | 242 |
| | | | | Early | | 248 |
| | Paleozoic | Permian | | Late | | 256 |
| | | | | Early | | 290 |
| | | Pennsylvanian | | | | 323 |
| | | Mississippian | | | | 354 |
| | | Devonian | | Late | | 370 |
| | | | | Middle | | 391 |
| | | | | Early | | 417 |
| | | Silurian | | Late | | 423 |
| | | | | Early | | 443 |
| | | Ordovician | | Late | | 458 |
| | | | | Middle | | 470 |
| | | | | Early | | 490 |
| | | Cambrian | | D | | 500 |
| | | | | C | | 512 |
| | | | | B | | 520 |
| | | | | A | | 543 |
| Precambrian | Proterozoic | Late | | | | 900 |
| | | Middle | | | | 1600 |
| | | Early | | | | 2500 |
| | Archean | Late | | | | 3000 |
| | | Middle | | | | 3400 |
| | | Early | | | | 3800? |

Figure: Geologic time scale courtesy of USGS

**SKILL 22.2**      **Apply the laws and principles of geology (e.g., law of original horizontality, law of superposition) to interpret diagrams of rock strata.**

**Principle of Cross-Cutting Relations**: A rock is younger than any rock it cuts across.

**Principle of Original Horizontality**: Rock units are originally laid down flat. Something happened to cause them to change orientation.

**Principle of Super Position**: The rock on the bottom is older than the rock on top.

**Principle of Biologic Succession**: Fossils correspond to particular periods of time.

Fossils appear and disappear periodically, providing a geologic time yardstick.

Geologists interpret the order of physical events or the geologic history of an area by observing and creating standardized diagrams of rocks. Because, in most areas, rock formations are covered by vegetation or development, or are underground, geologists must look in certain places to find rock formations that are uncovered and show deeper rocks. These include roadcuts and areas where rock faces have been exposed through weathering. An alternative is drilling into the rock and bringing a core to the surface to be examined.

From these sources of information, a **stratigraphic column** is created:

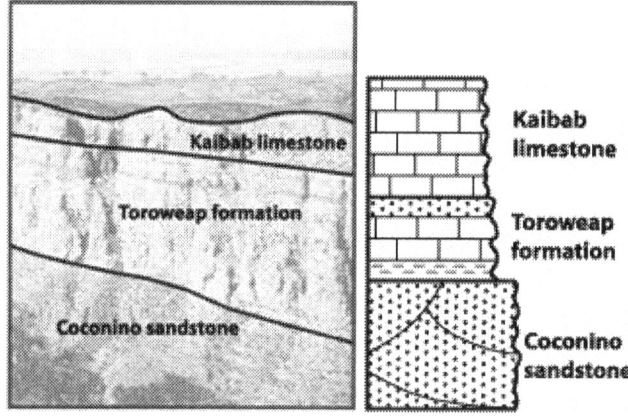

In the diagram above, rocks exposed in the Grand Canyon are translated into a stratigraphic column. This is a fairly simple column, as the most recent rocks are at the top, with the older rocks on the bottom. Because there are a lot of rocks exposed in this area of the western United States, these rock formations have been well-described and are easily recognized by geologists in the field.

The stratigraphic column contains various types of information. In this case, the rock formations are actually named. This allows this rock face to be compared to other exposed rocks distant from this site, which provides a large-scale understanding of the geology of the region. In some cases, such as a stratigraphic column for a core sample in a new area, the rocks would simply be described rather than named. In that case, they would be given names based on the type of rock, such as sandstone, shale, slate, limestone, granite, clay, etc.

Certain universal symbols are used to draw different kinds of rock layers, based on structural characteristics. For example, sandstones are drawn using dots, and limestones are drawn as bricks. From the diagram above, you can see that the Coconino sandstone has also been drawn with additional lines through it, to indicate fracturing. This can be important to geologists looking for water, oil, or natural gas deposits. A thin layer of shale is shown between the Coconino and Toroweap formations as a series of dashed lines.

In other formations, a silty shale or interbedded silts and shales might be shown as dashed lines mixed with dots. Salts may be shown using crosses, and basalts may be shown as black areas, sometimes with vertical lines if they are columnar basalts. Metamorphic rocks may be shown with wavy lines indicating folds; granite rocks as various solid colors with crosses or dots in them. Although there are certain conventions that are typically followed, each area will have its own specific formations and details that will be noted in the legend of a stratigraphic column. Colors are frequently used to indicate the actual color of the rock. Some examples of stratigraphic columns can be found in the links below:

http://www.igwa.org/iowacol.asp
http://www.deq.state.mi.us/documents/deq-gsd-info-geology-Stratigraphic.pdf

These columns also show the geologic era in which each layer was formed and the thickness of the layer, as well as shorthand letter notations for the layers.

Not all rocks appear from bottom to top, oldest to youngest. Originally they may be deposited this way, but then they can be subjected to a variety of geologic processes, including tilting, folding, uplifting, weathering, fracturing, and upwelling of magma within existing formations. These features can often be seen in roadcuts or other exposed surfaces and can confuse the interpretation of geologic events. Certain features are used to evaluate the order in which events occurred:

- **Rock type/appearance:** First, one should simply look carefully at each layer of rock and evaluate the type of rock, grain size, minerals, color, weathering, layering, and any other identifying characteristics that may be present. These can be compared to stratigraphic records for the area to identify the layer and its age.

- **Layering:** Geologists know that originally, the oldest layers of rocks were laid down on the bottom. This is true whether the rocks were formed through sedimentary processes (e.g., successive layers formed by weathering and erosion) or igneous processes (e.g., layers of lava laid down on top of one another). Due to gravity, these layers are normally laid down in a horizontal manner – i.e., the surface of the rock layers was originally horizontal.

- **Tilting:** Later, tilting of the rocks may have occurred due to processes related to plate tectonics. Generally, it is relatively rare to find rock layers that are tilted more than a 90° angle. Therefore, one can normally assume that the layer on the bottom is still the oldest, even if the layers are tilted.

- **Dating:** Certain types of rock layers can be dated through a number of means. Fossils in the rocks can be compared to known ages of fossils to determine the period in which those organisms lived, to date the rocks. Certain fossils represent organisms that lived only for a short period of time, and these are especially helpful in pinning down the timeframe in which a rock layer was deposited. Radioisotope dating, using a variety of different isotopes depending on the age of the rock layer being dated, can also be used to establish the relative ages of rock formations.

- **Intrusions:** Once layers of rock have been laid down, magma can intrude into the layer, either in large pockets known as batholiths, along fractures or in straight lines known as dikes. The age of these features is interpreted relative to the layers around them. If they cross other layers, then the intrusion is newer than those layers. The image below shows dark vertical dikes crossing nearly horizontal layers of lighter-colored rock in Alaska. Vertical dikes are sometimes called pipes.

- **Weathering:** Weathering can often differentially erode softer layers of rock without eroding harder layers. For example, sedimentary rocks may be eroded before harder igneous rocks. After a long period of weathering, this may leave features such as pipes, dikes, and batholiths exposed while the surrounding rock has completely weathered away. Half-Dome in Yosemite, CA (shown next) is a good example of a batholith that has been exposed by weathering; its round shape indicates its previous history as a magma intrusion.

Subsequently, additional sedimentary or igneous layers may be laid down around the exposed rock. Unfortunately, this can look similar to the case in which a body of magma intrudes into existing layers. Sometimes the margins of the intrusive layer can be examined to evaluate whether they were ever exposed to the elements (which would indicate that the layers surrounding them are newer) or to see if there is evidence of chemical interaction and/or melting of the surrounding rock (which would indicate that the layers surrounding them are pre-existing). In very complicated situations, the various rock layers may need to be compared among multiple locations to be properly interpreted.

**SKILL 22.3      Identify the principles, applications, and limits of radioactive dating.**

**Radiometric Dating**: The **most accurate** method of absolute dating, this technique measures the decay of naturally occurring radioactive isotopes. These isotopes are great timekeepers because their rate of decay is constant. Elements decay because of the inherent structure of the nucleus of the atoms. Neutrons hold the positively charged protons together. However, the positive protons attempt to repel each other. In some heavy elements, the protons repel each other to such a degree that the proton tears itself apart (decays) and by losing protons, becomes another element. The decay starts the moment an isotope crystallizes in a rock unit, and chemicals, weathering, environment, or temperature does not affect the rate of decay.

The radioactive decay causes the mother element to change into a daughter element. The Mother-Daughter relationship of produced nuclides during the series of isotope decay is the basis for radiometric dating. Although many isotopes are used in radiometric dating, the most widely known method is referred to as **Carbon-14 dating**. Carbon-14 is unstable and decays, decomposes, and transmutes to Carbon-12. The dating process compares the ratio of Carbon-14 to Carbon-12 in an object. Since the decay occurs at a known rate, it is very predictable and can be used as a clock standard. However, Carbon-14 decays quickly and can only be used to date organic compounds less than 40,000 years old.

Knowing the **Half-Life** of the isotopes is the key factor in the radiometric dating process. If we know the half-life, we can compare the ratio of isotopes found in the object, and count backward to get an accurate date. The most common element checked is the ratio of Uranium to Lead. Example: 1 gram of 238Uranium. After 100 million years, you have 0.013g of 206Pb (lead) and 0.989 of 238U. After 4.5 bilion years, you have 0.433g 206Pb and .500g of 238U. Therefore, the half-life of 238U is roughly 4.5 billion years.

Note: Only Carbon-14 can be used to date organic compounds. The other isotopes are not found in organic compounds.

**Radiometric Isotopes Commonly Used in Absolute Dating**

| Mother Isotope | Daughter Isotope | Half-life (in years) | Dating Range (in years) |
| --- | --- | --- | --- |
| 40K (Potassium) | 10Ar (Aragon) | 1.3 billion | >10 million |
| 87 Rb (Rubidium) | 87St (Strontium) | 47 billion | >100 million |
| 235U (Uranium) | 407U (Uranium) | 700 million | >1 million |
| 14C (Carbon) | 206 Pb (Lead) | 5,730 | >750<40,000 |

SKILL 22.4   Use paleontological information to infer to geologic history of a given area.

**Theories on the origin of life**

The origin of life and the environment from which life has formed is widely studied and theorized. Oparin developed the idea of a "primordial soup." He stated the coacervates, or organic material that resembles a lipid, can form out of an oxygen free environment through exposure to sunlight. These coacervates are not cells in and of themselves, but through "fusion" may develop into primitive cells. Generally it is accepted that amino acids can form from pre-biotic conditions. The amino acids can combine to form a lipid by-layer that is similar to the cell membrane. There is polymerization of nucleotides and RNA is developed. Ribosomes are formed and protein synthesis begins.

## Fossils

Fossils are remains of organic material found in the bedrock. Fossils may be difficult to find as there are many Earth processes that destroy organic remains, such as metamorphism and igneous activity. Many of the fossils that are found are exoskeletons, shells, and bones. Soft tissue fossils are extremely rare. There are two major types of fossils: fossilized bodies (such as bones claws, teeth, and shells) and trace fossils or ichnofossils (footprints, nests, and dung).

## Methods of fossilization

Carbonization: The process of compression and compaction of organic materials in anaerobic environment.

Mold: An impression of the organism in sediment. External mold is an impression of the outside of a shell; an internal mold is the impression of an inside of a shell. After the impression is made, the organism dissolves.

Cast: An organism dissolves leaving an empty void in the sediment. These cavities are filled forming a natural cast.

Permineralization: Minerals infiltrate into pores in the hard material of the organism (i.e. shell or bone). The mineral preserves the structure.

Replacement: The hard material from the structure is dissolved and replaced with other minerals. The size and structure are imperfectly preserved.

## Use of the fossil record

Scientists use the fossil record in order to determine past environments at specific locations. Through the study of paleobotnay, paleoclimatology, and paleoecology, scientists can create a succession for a specific area or can create a timeline for Earth's history. It is also helpful in determining how past climates have affected organisms so that we can determine how climate change might affect today's organisms.

**COMPETENCY 23.0**     Understand the hydrologic cycle and the processes by which water moves on and beneath the Earth's surface, and use this knowledge to analyze local weather budgets.

**SKILL 23.1**     Analyze a cross-sectional diagram of a water table and surrounding rock strata to predict movement of groundwater (e.g., aquifers, gradient).

## Groundwater

Groundwater is usually at a constant temperature and free of contaminants and suspended load. Only minimal treatment is required to remove man-induced toxins and lead. Groundwater is generally chemically uniform and classified as hardwater. **Hardwater**: water that has a high concentration of calcium, magnesium, and iron. Hardwater leaves red rings and a hard film because of its mineral deposits. Red deposits are from iron. Soap is not as sudsy in hardwater as it normally is. Hardwater can be turned into **softwater** through the use of sodium in equipment designed to exchange sodium for the calcium, magnesium, and iron. Groundwater flows and the path taken by it somewhat mirrors the topography of the area.

## Groundwater Zones

**Zone of Aeration (Vadose Zone):** the underground groundwater zone filled with both water and air.

**Zone of Saturation**: the underground groundwater spaces totally filled with water.

**Water Table**: the water table marks the top of the Zone of Saturation, and roughly follows the topography.

**Aquifer**: the porous and permeable layer of rock or sediment through which the groundwater moves freely. An aquifer can come at any depth. The type of material the water is passing through determines the rate of flow in the aquifer. Depending on the composition of the soil, pockets of aquifers can form. Example: Water passes easily through Limestone, while water passes slowly through sand but sand is still permeable and porous for forming an aquifer. Silts and clays are porous, but non-permeable and become Aquitards.

**Aquitards**: non-pourous, non-permeable layers that trap water in an aquifer.

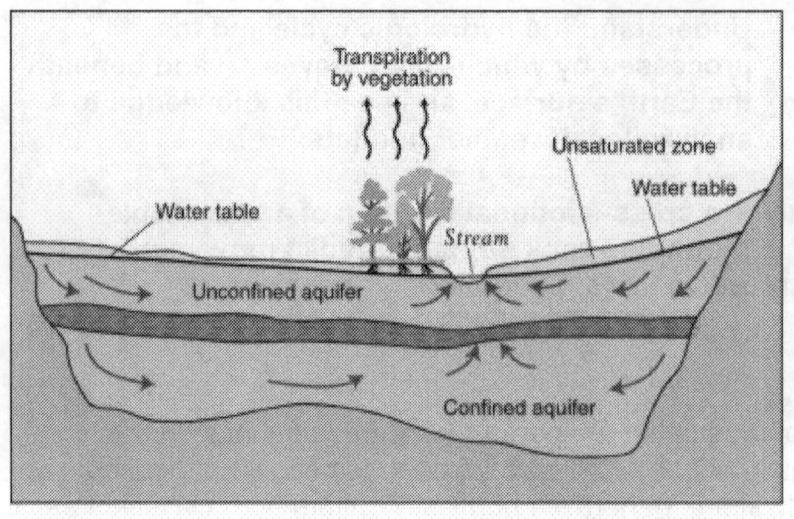

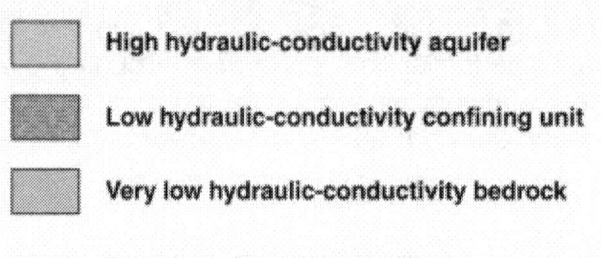

**SKILL 23.2** Describe the effects of various factors (e.g., vegetation, gradient, rock strata) on components of a local water budget.

In sediments or soils, **porosity** refers to the amount of void or open space between grains. It is usually represented as a percentage or fraction from 0-1. In very loose sediments with large, round grains, porosity can be as high as 50%. In poorly sorted sediments with a variety of grain sizes, smaller particles fill the voids and porosity can range down to 20-30%. Rocks have much lower porosity, with sedimentary rocks being more porous than granitic rocks, which may have a porosity as low as 0.01 (1%). Fractures in the rock may also increase porosity and affect water flow in unpredictable ways. This is known as **secondary porosity**.

**Permeability** is the ability of sediments or soils to allow the flow of water or another liquid such as oil. Permeability is related to porosity, but also depends on the size and connectedness of the pore spaces. Permeability is expressed in units of $cm^2$. Generally, sandy or gravelly soils are much more permeable than clays or silts, even though clays and silts have a relatively high porosity. In clays and silts the pore spaces are so small that it is hard for water to pass through them. Water has a high **surface tension**, and therefore it cannot penetrate into very small spaces.

Permeability values of $10^{-3}$ to $10^{-6}$ $cm^2$ are considered to be high-medium permeability and make good groundwater aquifers (often sands and gravels or highly fractured rock), while permeability values of $10^{-11}$ $cm^2$ or smaller (often silts and clays or unfractured rock) are considered impermeable and are barriers to flow. Permeability is measured in the laboratory or the field using a pump test.

**Water pressure** is the height at which water will rise in a well, or **hydraulic head**. The height of water in an unconfined aquifer is known as the **water table**, which may fluctuate with rainfall, tides, or human extraction. Aquifers confined by a low-permeability layer may contain water under higher pressure. If a well is drilled into such an aquifer, the water may rise higher than the confining layer.

These various properties are combined into **Darcy's Law**, which predicts the rate of discharge of water through soils or sediment:

$$Q = \frac{-\kappa A}{\mu} \frac{(P_b - P_a)}{L}$$

where:  $Q$ = discharge volume of water ($cm^3$/sec)
$\kappa$ = intrinsic permeability of soil or sediment ($cm^2$)
$A$ = cross-sectional area through which the water flows ($cm^2$)
$P_b - P_a$ = water pressure change over the distance $L$ (Pa)
$\mu$ = viscosity of water (Pa-sec)
$L$ = distance (cm)

This equation is also frequently written as:

$$q = -K(\Delta h/\Delta l)$$

where:  $q$ = Darcy flow velocity (cm/sec)
$K$ = hydraulic conductivity (cm/sec)
$\Delta h$ = change in water pressure or hydraulic head (m)
$\Delta l$ = distance over which water travels (m)

The quantity ($\Delta h/\Delta l$), also notated as $i$, is known as the **hydraulic gradient**, or the change in hydraulic head over distance. Notice that water only flows from higher to lower hydraulic head, which is why the sign of the equation is negative. The greater the hydraulic gradient, the greater the flow will be.

The Darcy flow velocity $q$ assumes that water takes a straight path through the aquifer. However, water actually has to navigate through the pore spaces and because of this, takes an indirect path through the sediments. The actual velocity of the water required to navigate through the pore spaces and create the Darcy flow velocity $q$ is called the **mean porewater velocity**, and is higher than the Darcy velocity. This quantity is given by the following equation:

$$v = q/\phi$$

where:  $v$ = mean porewater velocity (cm/sec)
$q$ = Darcy flow velocity
$\phi$ = porosity

**Example:** What is the mean porewater velocity in a sandy aquifer with a hydraulic conductivity of 1.5 cm/sec and a porosity of 0.43, if the hydraulic head decreases by 10 m over a distance of 1 km?

**Solution:** First, use the flow version of Darcy's law to calculate the Darcy flow velocity:

$$q = -1.5 \text{ cm/sec} \times -10 \text{ m} / 1000 \text{ m} = 0.015 \text{ cm/sec}$$

Next, use the equation for mean porewater velocity to find the solution:

$v = 0.015$ cm/sec $/ 0.43 = 0.035$ cm/sec

# TEACHER CERTIFICATION STUDY GUIDE

**COMPETENCY 24.0**     **Understand ocean water and its movements.**

**SKILL 24.1**     **Describe the circulation patterns in the oceans and factors that influence these patterns (e.g., temperature variations, wind systems, Coriolis effect).**

## WIND SYSTEMS

The movement of ocean water is caused by the wind, the Sun's heat energy, the Earth's rotation, the Moon's gravitational pull on Earth, and by underwater earthquakes. Most ocean waves are caused by the impact of winds. Wind blowing over the surface of the ocean transfers energy (friction) to the water and causes waves to form. Waves are also formed by seismic activity on the ocean floor. A wave formed by an earthquake is called a seismic sea wave. These powerful waves can be very destructive, with wave heights increasing to 30 meters or more near the shore.

## TEMPERATURE VARIATIONS

In the world's oceans, thermohaline circulation is coupled with and driven by surface water circulation, as shown in the diagram below.

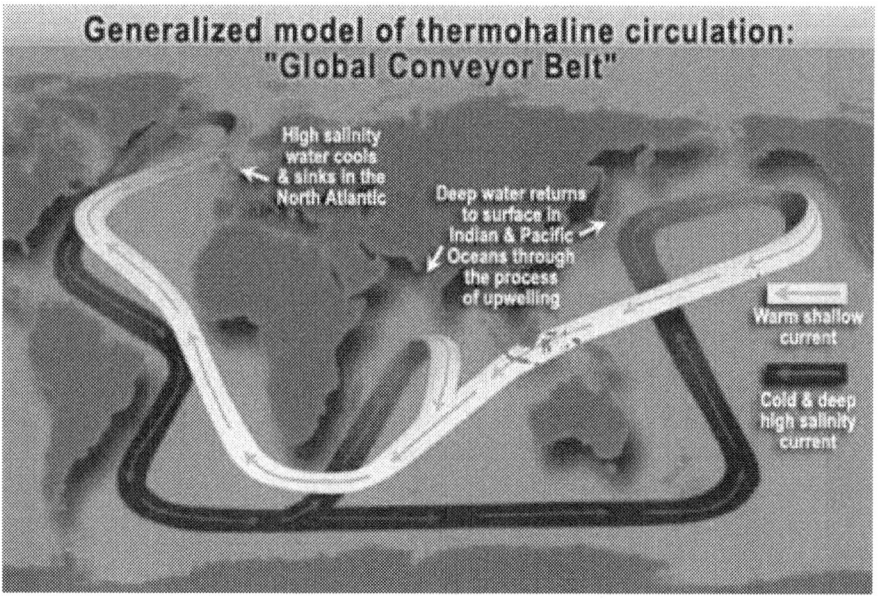

The term **thermohaline** refers to temperature "thermo" and salinity "haline", the two factors that drive this circulation pattern. As warmer surface water, known as the Gulf Stream, travels up the coast of eastern North America, it eventually meets cold dry air from Canada. This continental air cools and evaporates some of the surface water, leaving it colder and more saline than before. The formation of polar ice also removes more freshwater, making the ocean water even saltier. This dense, cold ocean water sinks and travels back down along the east side of the North American continent, driving the overall circulation patterns.

Patterns of deep ocean circulation are affected in part by the shapes of the continents, by the need to maintain a water balance in different parts of the ocean, and by areas of upwelling. Surface water sinks down into the deep ocean at the top of the Gulf Stream, and then travels away from that area along North and South America, and subsequently along Antarctica. Deep waters return to the surface in areas of upwelling; these occur in the Indian Ocean, in the North Pacific offshore of Oregon, and in the South Pacific offshore of Peru and Chile. Areas of upwelling are very rich in nutrients and are excellent fisheries resources. Water requires approximately 1000 years to travel along the entire circulation pattern.

Thermohaline circulation is very important to regulating climate and vice versa. Global warming can slow down thermohaline circulation through the melting of sea ice and glaciers at the poles. When large amounts of freshwater are added at the poles, it counteracts the evaporation that normally occurs and seawater does not sink, because it does not become more saline or denser. It has been hypothesized (and there is evidence in the geologic record) that this could stop global thermohaline circulation altogether.

Thermohaline circulation strongly affects the climate of the continents through the movement of warm surface waters. For example, if this circulation were to stop, the Gulf Stream would not travel up the eastern United States and across to the United Kingdom, and it would be much colder in these areas than it is now. The ultimate effects of such changes are difficult to fully predict, but it is certain they would have significant impacts on local climate, agriculture, and fisheries resources.

**Major Ocean Circulation Currents**

The primary currents in both hemispheres are the **Equatorial Counter Current.** These currents are located slightly above the equator in the northern hemisphere and slightly below the equator in the southern hemisphere. They are offset from the equator because of the ITCZ (Inter Tropic Convergence Zone).

## Western Boundary Currents

All six of the major Western Boundary Currents flow north.

**Gulf Stream** (also called the Florida Current)- Atlantic Ocean
The Gulf Stream receives more water than it theoretically should. This is because the Brazil Current supplements it. This extra water is present because of the shape of Brazil; it forces the water up into the lower part of the Gulf Stream in the Caribbean.

**Brazil Current** (Southern Hemisphere equal to the Gulf Stream)- Atlantic Ocean

**Japan Current** (Kuroshio Current)- Pacific Ocean

**East Australian Current**- Pacific Ocean

**Agulhas Current** - Indian Ocean

**Somalia Current** -Indian Ocean
Note: the Somalia current is only present one half of the year. The rest of the time it flows south.

Because the Pacific Ocean is so large, the Western Boundary Currents in the north and south Pacific (Kuroshio Current and East Australian Current) each have two Gyres. Additionally, the Kuroshio Current is a big player in the El Nino effect.

Antarctic water freely circulates causing the southern ocean to remain fairly constant in temperature. The West Wind Drift causes the Antarctic flow.

In the Indian Ocean, the southern equatorial current is affected by the Southwest Monsoon winds. When the SW Monsoon winds blow, it reverses the current of the northern leg of the South Equatorial current. That leg is renamed and becomes the SW Monsoon Current.

## Directions

When trying to figure out the direction of a current or air stream use the following:

**Air circulation**: The direction indicated by the name is coming from that direction. Example: North Westerlies are from the northwest.

**Ocean circulation**: The direction indicated by the name is going to that direction. Example: The Northwest current is going to the northwest.
*Rule of Thumb: Fly from, Sail to.*

**Vorticity** is highly connected with ocean motion. It affects which way the current is going to flow.

**Cold Core Eddies**: Rotate counterclockwise.
**Warm Core Eddies**: Rotate clockwise.
*Reverse the rotation in the Southern Hemisphere.*

Locating the warm and cold water eddies is simple:

Southside of the Gulf Stream = Cold core eddy.
Northside of the Gulf Stream = Warm core eddy.

**SKILL 24.2**     **Describe the types, causes, and effects of tidal and wave motions of ocean water.**

### Waves

A transfer of energy from tides, currents, or wind causes waves. Water will remain in place unless moved by the current or tide. Currents are caused by changes in water density, salinity, and pressure. Tides are primarily caused by the gravitational pull of the Moon.

Waves move in an orbital pattern, causing an up and down motion. They have a forward or lateral motion only if moved by the wind, current, or tides. The depth of the waveform where the energy is felt is equal to ½ of the wavelength. Below that depth, the water remains relatively calm.

When a wave approaches the shore, the circular orbit action flattens out and becomes more elliptical. As the wavelength shortens, the wave steepens until it finally breaks, creating surf. The waves break at a distance of 1/20 of the wavelength.

### Tides

The periodic rise and fall of the liquid bodies on Earth are the direct result of the gravitational influence of the Moon and to a much lesser extent, the Sun.

Tides are produced by the differences between gravitational forces acting on parts of an object. As shown in Netwon's Universal Law of Gravitation, the gravitational effect of two bodies is mutually constant and depends largely on the distance and mass between the objects.

The side of the Earth that faces the Moon is roughly 4,000 miles (6,400 km) closer to the Moon than is the Earth's center. This has the effect of increasing the Moon's gravitational attraction on Earth's oceans and landforms. Although the effect is so small on the mass of the landforms as to be invisible, the effect on the liquid parts is greater.

The Moon's gravitational effect causes a bulge to form on both sides of the Earth. If we were able to view such subtle change from outer space, the affected waters would create an elliptical shape, compressing downward at the top and bottom of the planet and extending outward on the sides.

This double-bulge effect causes the tides to fall and rise twice a day, and the time of the high and low tides is dependent on the phase of the moon.

Yet not all locations are uniformly affected. The tidal cycle at a particular location is actually a very complicated interaction of the location's latitude, shape of the shore, etc.

Example: The Bay of Fundy has a twice-daily tide that exceeds 12 meters, while the northern coast of the Gulf of Mexico only has one tidal cycle that seldom exceeds 30 centimeters rise and fall.

Because of its distance from the Earth, the Sun's gravitational effect on tides is only half that of the Moon's.

However, when the gravitational effects of both the Sun and Moon join together during a new moon and a full moon phase, the tidal effects can be extreme.

During a new moon and a full moon, tidal effects are much more pronounced as the tidal bulges join together to produce very high and very low tides.

These pronounced tides are collectively known as Spring Tides. During the first and third quarters of the moon phases, the Sun's effect is negligible and consequently, the tides are lower. These are Neap Tides.

**COMPETENCY 25.0**     **Understand the structure and topography of the ocean basin.**

**SKILL 25.1**     **Identify ocean zones (e.g., littoral, benthic) in terms of their physical characteristics.**

The ocean is commonly divided into five layers according to depth. Each layer has its own characteristics and a specific group of organisms reside there. The depth of water is key because photosynthesis requires solar energy. Therefore, the uppermost layer, where the light can penetrate, is where we find phytoplankton. Phytoplankton are small, photosynthetic organisms which are the base of the oceanic food chain.

Epipelagic Zone – This layer extends from the surface to 200 meters (656 feet). It is in this zone that most of the visible light exists. The majority of plankton and fish are found here, as well as their predators (large fish, sharks, and rays).

Mesopelagic Zone - Extending from 200 meters (656 feet) to 1000 meters (3281 feet), the mesopelagic zone is also referred to as the twilight or midwater zone. Very little light penetrates here. Instead, most of the light observed is generated by bioluminescent creatures. A great diversity of strange fishes can be found here.

Bathypelagic Zone – This layer extends from 1000 meters (3281 feet) down to 4000 meters (13,124 feet). Here there is no penetration by solar light, so any light seen is in the form of bioluminescence. Most of the animals that live at these depths are black or red in color due to the lack of light. The water pressure at this depth is quite large, but a surprising number of creatures can be found here. Common inhabitants include some fish, molluscs, jellies, and crustaceans. Sperm whales can dive down to this level in search of food.

Abyssopelagic Zone - Extending from 4000 meters (13,124 feet) to 6000 meters (19,686 feet), this zone has the least inhabitants. The water temperature is near freezing, and there is no light at all. Common organisms include invertebrates such as basket stars and squids. The name of this zone comes from the Greek meaning "no bottom" and refers to the ancient belief that the open ocean was bottomless.

Hadalpelagic Zone - This layer extends from 6000 meters (19,686 feet) to 10,000 meters (32,810 feet)- the sea floor. These areas are most often found in deep water trenches and canyons. In spite of the unimaginable pressures and cold temperatures, life can be found here. Generally, these include life forms that tolerate cool temperatures and low oxygen levels, such as starfish and tubeworms. The exception to this rule would be chemosynthetic communities living near deep-sea vents. These creatures create their own nutrients from carbon dioxide or methane released by the hot thermal vents. Chemosynthetic organisms then become prey to larger organisms. As such, chemosynthetic organisms are also primary producers and are at the bottom of the food chain, just like their photosynthetic friends, although they are at the opposite end of the ocean!

**SKILL 25.2**     **Describe the major structural features of the ocean floor.**

Four major features are found on the sea floor: Continental Margins, Ocean Floor, Ridges, and Trenches.

**Continental Margins**: where the shore and water meet. There are three types of continental margins: the Continental Shelf, Continental Slope, and Continental Rise.

**Continental Shelf**: the low slope of land immediately offshore in the ocean. Contains sediment carried off the continents. The sediment buildups the continental shelf to variable widths, and can be cut by canyons known as Submarine Canyons.

**Continental Slope**: the point where the sediment tumbles into the deep ocean. Much steeper slope than the continental shelf, it extends from the edge of the shelf to the deep ocean floor. Material is tumbled down the slope by Turbidity Currents. However, the most prominent features are underwater avalanches. Sediment in the area is very fine and unstable, and it falls off and moves very rapidly. In effect, parts of the continental shelf fall down to the ocean bottom.

**Continental Rise**: the mound of sediment at the bottom of the continental slope. Formed from the material tumbled off the continental shelf by the turbidity currents.

**The Ocean Floor**

The ocean floor starts at the end of the continental rise and has an extensive Abyssal Plain. It is a flat, featureless landscape, broken only by ridges or trenches. Covered with deep deposits of sediment, the ocean floor is the flattest topography on Earth.

### Ocean Floor Sediment Deposits

**Terrigenous**: deposits derived from the land. Includes very fine sediments from run off and erosion, and some sand, ash, and sediment carried by the wind in sand storms and volcanic events.

**Glacial Deposits**: material dropped out of an iceberg as it melts (uncommon event).

**Biogenous**: predominant sediment of the Abysmal Plain. Derived from biologic organisms, mainly carbonate and silica shells of microscopic plankton. When plankton die their shells drop to the bottom and accumulate.

**Hydrogenous**: material that precipitates out of the seawater. The two most common chemical nodules formed from solution are Manganese and Phosphate nodules. Both are a viable commercial ore source found in localized deposits.

### Ridges

There are two types of ridges, mid-ocean and aseismic.

**Mid-Ocean Ridge**: Characterized by great length and steep slopes, this is the site of sea floor spreading.

**Aseismic Ridge**: Formed by hot spots, islands, sea mounts, and guyots.

**Sea Mount**: a volcano that never reached the surface. A sea mount is a pointy topped undersea mountain usually formed by a hot spot as part of a ridge or chain.

**Guyot**: a flat-topped undersea mountain. Originally the mountain was above the Surface. Either erosion and/or subsidence of sea floor bulging when the tectonic plate moved off of a hot spot caused it to sink beneath the ocean surface. There is a chain of aseismic guyots linking Midway Island to Hawaii.

**Trenches**: The deepest parts of the ocean. They break the abysmal plain, have steep slopes, and are found at the sites of subduction zones.

## Classification of Continental Margins

There are three types of continental margins and they are classified by tectonic setting.

**Passive Margin**: an area of little activity, no subduction, and no Earthquakes. A spreading zone usually borders a passive margin. Example: Atlantic Coast of the U.S. A passive margin is typically characterized by very large continental shelf/slope/rise. It also has prominent features such as Drowned River Valleys.

**Drowned River Valleys**: When the sea level was lower, river valleys were cut in the normal fashion. Later, the sea level rose, covering the valleys. Example: The Chesapeake Bay is a series of drowned river valleys. This accounts for the fingered look of its meandering inlets leading to a river mouth.

**Active Margin**: an area of active tectonics.

**Subduction Boundary**: a narrow continental shelf bordered by a deep trench. Areas similar to the Pacific NW along the Washington and Oregon coastlines.

**Transform Boundary**: Areas similar to the California coastline. Narrow continental shelf, but the shelf is cut by a series of submarine canyons that run parallel to each other but perpendicular to the shoreline. Shaking along the Earthquake fault lines causes the submarine canyons to form. Example: Monterey Bay is actually a large submarine canyon.

**SKILL 25.3**     **Identify the factors involved in changing the structure of the ocean floor.**

**Oceanic crust** is the rocky part of the Earth's surface that lies beneath the ocean and forms a surficial crust above the mantle. Oceanic crust is always being destroyed and regenerated through tectonics.

- **Composition:** Oceanic crust is made up of **mafic** rock, which has a relatively high concentration of heavier elements such as magnesium and iron. It has a lower concentration of silica than continental crust.

- **Density:** Because of its composition, oceanic crust is denser than continental crust. When the two come into contact at the continental margins, oceanic crust tends to get **subducted**, or pulled beneath, the continental crust due to its greater density.

- **Color:** Mafic oceanic rock is generally dark or black in color.

- **Thickness:** Oceanic crust is relatively thin, from 7-10 km thick (see figure).

- **Age:** Oceanic crust is constantly being created at mid-ocean ridges, where magma rises to the surface and spreads out. Oceanic crust is also constantly being destroyed at subduction zones along the edges of continents or where two oceanic plates meet, at which time it is reincorporated into the mantle. Because oceanic crust is constantly being created and destroyed, its maximum age is about 200 million years. Continental crust, on the other hand, is generally much older.

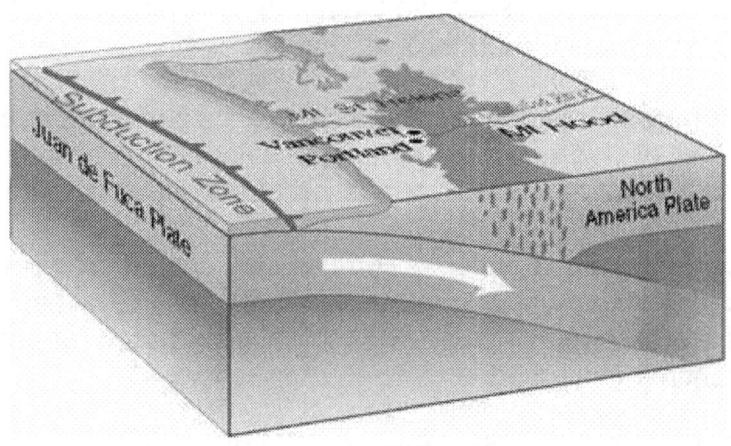

**COMPETENCY 26.0**       Understand marine life and the marine habitat.

**SKILL 26.1**       Identify the characteristics and major groups of marine plants and animals.

### Marine Plant Life (Kingdom Plantae)

Marine plant life can be divided into the following three major categories: Sea grasses, mangroves, and algae.

### Sea grass

Sea grasses are flowering marine saline environment plants from the plant families Posidoniaceae, Zosteraceae, Hydrocharitaceae, and Cymodoceaceae. Sea grasses rely on photosynthesis to generate their own energy and are found in the photic zone. Sea grasses grow in shallow, sheltered coastal waters anchored in sand or mud bottoms, and can form extensive beds or meadows. These plants are named for their resemblance to terrestrial grasses of the Family Poaceae. Sea grasses complete their entire life cycle, including pollination, completely submerged.

### Mangroves

Mangroves are trees and shrubs of tropical saline coastal habitats. The term mangrove may be used to refer specifically to the mangrove family of plants, the Rhizophoraceae, constituted by tropical or subtropical flowering plants. These are woody plants with opposite or whorled, above-water leaves and extensive below-water root systems. These plants have insect-pollinated flowers with five petals and a nectar-containing disc. Mangrove species may be hermaphroditic and viviparous – producing seeds that germinate before becoming detached from the parent plant. Approximately 110 species have been identified as belonging to the plant assemblage mangal. Each of these species has developed a set of physiological adaptations to overcome the problems of anoxia, salinity, and tidal inundation, though tolerances to these conditions may differ among mangrove species.

## Algae

The term algae may be used to describe several groups of photoautotrophic aquatic organisms that span the kingdoms of plant and protist. These plants produce energy through cynobacterial photosynthesis and use it to convert inorganic substances to organic matter. While most types of algae are single-celled and microscopic, multi-cellular species, such as giant kelp, also exist. Multi-celled algae with differentiated form may be called seaweed. All forms of algae have simple reproductive systems and lack the leaves, roots, flowers, seeds, and other organ structures of vascular plants. Microscopic forms of algae that live suspended in the water column are called phytoplankton, and provide the food base for most marine food chains. All forms of algae are found within the photic zone.

## Marine Animal Life (Kingdom Animalia)

The marine animal kingdom, like the terrestrial animal kingdom, is divided into vertebrates and invertebrates.

## Marine Invertebrates

Marine invertebrate species make up a huge percentage of aquatic life. Like terrestrial invertebrates, a marine invertebrate is any animal that lacks a spinal column. Some species of marine invertebrate live only below water, while others may be capable of moving between terrestrial and aquatic environments. Marine invertebrates include anemones, crabs, fan worms and feather dusters (annelids), jellyfish, lobsters, scallops and oysters, sea cucumbers, sea slugs and nudibranchs, sea urchins, shrimp, snails, sponges, starfish, and tridacna clams.

## Marine Vertebrates

### Fish

Fish differ from other organisms in their distinct anatomies. Fish have two-chambered hearts, operculums, secretory cells that produce mucous, swim bladders, scales, fins, lips, and eyes. Fish breathe by extracting oxygen from water through their gills and use fins to move and stabilize themselves in water. Common fish types include sardines, anchovy, tuna, clownfish, and bottom fish.

### Marine Reptiles

Marine reptiles are reptiles (tetrapods and amniotes, whose embryos are surrounded by an amniotic membrane and members of the class Sauropsida) that have become secondarily adapted for an aquatic or semi-aquatic life in a marine environment. Most marine reptiles live on or near land. Common marine reptiles include sea turtles, marine iguanas, sea snakes, and crocodiles.

## Marine Birds

Birds are bipedal, warm-blooded, egg-laying vertebrate animals. Many of these species spend much of their time on, in, or near the ocean. Marine birds generally feed on organisms found in the ocean, and include penguins, albatross, gulls, pelicans, etc.

## Marine Mammals

Marine mammals are mammals that are primarily ocean-dwelling or that depend on the ocean for food. These animals differ from most other marine organisms in that they generally breathe air, have hair and a thick layer of blubber for insulation, give live birth, feed off milk as young, and maintain a high internal body temperature. The five main types of marine mammals include:

1. Order Sirenia: the manatee, dugong.
2. Order Carnivora, family Ursidae: the polar bear
3. Order Carnivora, superfamily Pinnipedia: the seal, sea lion, and walrus
4. Order Carnivora, family Mustelidae: the Sea Otter and Marine Otter
5. Order Cetacea: the whale, dolphin, and porpoise

**SKILL 26.2    Describe the zonation of marine plants and animals.**

Aquatic ecosystems are, as the name suggests, ecosystems located within bodies of water. Aquatic biomes are dived between fresh water and marine. Freshwater ecosystems are closely linked to terrestrial biomes. Lakes, ponds, rivers, streams, and swamplands are examples of freshwater biomes. Marine areas cover 75% of the earth. This biome is organized by the depth of the water. The intertidal zone is from the tide line to the edge of the water. The littoral zone is from the water's edge to the open sea. It includes coral reef habitats and is the most densely populated area of the marine biome. The open sea zone is divided into the epipelagic zone and the pelagic zone. The epipelagic zone receives more sunlight and has a larger number of species.

The ocean floor is called the benthic zone and is populated with bottom feeders. Marine biomes include coral reefs, estuaries, and several ocean systems.

## Oceans

Within the world's oceans, there are several separate zones, each with it's own temperature profiles and unique species. These zones include intertidal, pelagic, benthic, and abyssal. The interdidal and pelagic zones are further distinguished by the latitude at which they occur (species have evolved to live in the various temperature waters). The intertidal zone is the shore area, which is alternately under and above the water, depending on the tides. Algae, mollusks, snails, crabs, and seaweed are all found in the intertidal zones. The pelagic zone is further from land but near the surface of the ocean. This zone is sometimes called the euphotic zone. Temperatures are much cooler than in the intertidal zone and organisms in this zone include surface seaweed, plankton, various fish, whales, and dolphins. Further below the ocean's surface is the benthic zone, which is even colder and darker. Much seaweed is found in this zone as well as bacteria, fungi, sponges, anemones, sea stars, and some fishes. Deeper still is the abyssal zone, which is the coldest and darkest area of the ocean and has high pressure and low oxygen content. Thermal vents found in the abyssal zone support chemosyntheic bacteria, which are in turn eaten by invertebrates and fishes.

## Coral reefs

Coral reefs are located in warm, shallow water near large landmasses. The best known example is the Great Barrier Reef off the coast of Australia. The coral itself is the predominant life form in the reefs and obtains its nutrients largely through photosynthesis (performed by the algae). Many other animal life forms also populate coral reefs including many species of fish, octopuses, sea stars, and urchins.

## Estuaries

Estuaries are found where freshwater and seawater meet, for instance where rivers flow into the oceans. Many species have evolved to thrive in the unique salt concentrations that exist in estuaries. These species include marsh grasses, mangrove trees, oysters, crabs, and certain waterfowl.

## Ponds and Lakes

As with the other aquatic biomes, many varied ecosystems occur in ponds and lakes. This is not surprising since lakes vary in size and location. Some lakes are even seasonal, lasting just a few months each year. Additionally, within lakes there are zones, comparable to those in oceans. The littoral zone, located near the shore and at the top of the lake, is the warmest and lightest zone. Further from land, but still at the surface of the lake is the limnetic zone. Deeper in the lake is the profundal zone, which is cooler and darker.

## Rivers and Streams

This biome includes moving bodies of water. As expected, the organisms found within streams vary according to latitude and geological features. Additionally, characteristics of the stream change as it flows from its headwaters to the sea. Also, as the depth of rivers increases, zones similar to those seen in the ocean are seen. That is, different species live in the upper, sunlit areas (e.g., algae, top feeding fish, and aquatic insects) and in the darker, bottom areas (e.g., catfish, carp, and microbes).

## Wetlands

Wetlands are the only aquatic biome that is partially land-based. They are areas of standing water in which aquatic plants grow. These species, called hydrophytes are adapted to extremely humid and moist conditions and include lilies, cattails, sedges, cypress, and black spruce. Animal life in wetlands includes insects, amphibians, reptiles, many birds, and a few small mammals. Though wetlands are usually classified as a freshwater biome, they are in fact salt marshes that support shrimp, various fish, and grasses.

**SKILL 26.3** **Describe the relationships between marine organisms and the marine environment.**

## The Intertidal Zone

The Intertidal Zone is the area on the shore that is intermittently covered by tide water. The Intertidal Zone is the home of many marine organisms. Algae, sea grass, seaweed, scavengers, and certain crustaceans can be found in the Intertidal Zone. The more undisturbed a beach is, the greater the biodiversity of the area. Photoplankton, the most numerous primary producers on the planet, are abundant here. Photoplankton, also found in **the Epipelagic Zone of the ocean, provide most of the oxygen for the planet.**

## Reefs

Biodiversity in reefs is great. The calcium depositing animals provide shelter and protection for many fish and zoozanthellae. Coral usually grows on top of rocky outcrops, although they are able to grow on other medium as well, which has allowed humans to create artificial reefs. Reefs are crucial to overall health of the oceans as well as the planet. Life in the reefs is crucial to oxygen production for the planet. El Niño and ocean warming has been detrimental to the reefs. Over 58% of the reefs in the ocean are endangered due to rising ocean temperatures.

### The Open Ocean

The open ocean is defined as the area between points of land or seamounts. Many organisms can be found in the open ocean, but usually it is simply a pass through for organisms traveling from one place to another. Organisms such as tuna, dolphins, whales, sharks, and sea turtles can be found in the open ocean.

### Deep Sea Trenches

The Aphotic Zone, or the deep sea, is the zone where sunlight no longer reaches. Even though this area is dark and under high pressure, a great variety of life forms can be found. Many of these organisms are bioluminescent, or have the ability to create their own light. In addition, certain shrimp and small flounder can be found. Blue whales have been known to dive this deep to feed. Many organisms found in the deep sea will rise to seamounts to feed and spawn. In addition to organisms such as fish, microbes and bacteria can be found in abundance around hydrothermal vents or black smokers. These organisms are called chemotrophs, as they feed off of the sulfur and other chemicals spewed by these vents.

## Sample Test

Directions: The following are multiple choice questions. Select from each grouping the best answer.

**1. Which layer of the atmosphere would you expect most weather to occur?**

A. troposphere
B. thermosphere
C. mesosphere

**2. What percentage of earth's surface is covered by water?**

A. 61%
B. 71%
C. 81%

**3. Which layer of the earth's atmosphere contains the Ozone layer?**

A. thermosphere
B. troposphere
C. stratosphere

**4. Copernicus developed a theory that is known as**

A. baycenter
B. heliocentric
C. geocentric

**5. The boundary that separates the crust from the mantle is known as**

A. Moho
B. shadow zone
C. catacastic

**6. The product of intrusive activities would result in forming a**

A. cinder cone
B. volcanic pipe
C. dike

**7. A star's light and heat are produced by**

A. magnetism
B. electricity
C. nuclear fusion

**8. The center of an atom is called**

A. micron
B. nucleus
C. electron

**9. An important food source for animals in a water biome**

A. shrimp
B. plankton
C. seaweed

**10. The smallest piece of an element is called a/an**

A. compound
B. nucleus
C. atom

**11. An instrument that measures relative humidity is known as**

A. psychrometer
B. anemometer
C. barometer

12. Removing salts from ocean water by heating is called

A. filtration
B. distillation
C. freezing

13. When molecules in the air cool and combine to form rain, _____ has occurred.

A. condensation
B. convection
C. radiation

14. Which instrument measures wind direction?

A. anemometer
B. barometer
C. wind vane

15. The most important cause of erosion is

A. water
B. wind
C. air

16. An anemometer measures

A. wind velocity
B. temperature
C. relative humidity

17. The boundary that develops when a cold air mass meets a warm air mass

A. cold front
B. warm front
C. stationary front

18. When the sun, moon and earth are aligned in a straight line what type of tides are produced?

A. neap tides
B. high tides
C. spring tides

19. North of the equator, currents move in which direction?

A. counter-clock wise
B. clockwise
C. northerly

20. Rocks that serve as aquifers are

A. impermeable
B. permeable
C. igneous

21. Volcanoes with violent eruptions are known as

A. shield volcanoes
B. dome volcanoes
C. cinder volcanoes

22. The Richter scale measures

A. compressions
B. focus
C. magnitude

23. The earth's outer core is probably

A. liquid
B. solid
C. rock-bed

**24. What are the two most abundant elements found in the earth's crust?**

A. oxygen and oxides
B. oxygen and cabonates
C. oxygen and silicon

**25. The San Andreas Fault is classified as a**

A. transform fault
B. oblique-slip fault
C. reverse fault

**26. Batholiths are the largest structures of which type of rock activity?**

A. intrusive rock
B. extrusive rock
C. magma

**27. The main agents of chemical weathering are**

A. water, oxygen, $CO_2$
B. water, oxygen, sulfur
C. water, oxygen, nitrogen

**28. Soil classified as porous is called**

A. clay soil
B. laterites soil
C. sandy soil

**29. Intrusive igneous rock forms**

A. glassy texture
B. small crystals
C. large crystals

**30. Which types of rocks are rich sources of fossil remains'?**

A. sedimentary rock
B. metamorphic rock
C. intrusive rock

**31. The best preserved animal remains have been discovered in**

A. resin
B. lava
C. tar-pits

**32. The Mid-Atlantic is a major area of which type plate movement?**

A. subduction plate movement
B. divergent plate
C. convergent plate

**33. When lava cools quickly on the earth's surface the newly formed rock is called**

A. clastic
B. intrusive
C. extrusive

**34. When a dyke forms with magma flowing in a tub-like structure this is known to be a/an**

A. extrusive activity
B. intrusive activity
C. metaphoric activity

**35. A caldera is formed when a large depression collapses. This is the result of a**

A. sinkhole
B. aquifer
C. volcanic eruption

36. Trenches observed on the sea floor are the results of

A. interaction
B. divergence
C. subduction

37. Alfred Wegener's hypothesis of continental drift was not supported until scientists began studying

A. sea floor
B. mountain ranges
C. volcanoes

38. These massive waves are caused by the displacement of ocean water, and are often the result of underwater earthquakes.

A. epicenters
B. tidal waves
C. tsunamis

39. Sea floor spreading occurs when the earth's crust is stretched and pulled apart in a process called

A. slippage
B. rifting
C. drifting

40. The layer of the atmosphere that is in a plasma state and aids in communication

A. Thermosphere
B. Ionosphere
C. Mesosphere

41. A stream erodes bedrock by grinding sand and rock fragments against each other. This process is defined as:

A. dissolving
B. transportation
C. abrasion

42. Rocks formed from magma are:

A. igneous
B. metamorphic
C. sedimentary

43. Rocks formed by the intense heating or compression of pre-existing rocks are classified as

A. igneous
B. metamorphic
C. sedimentary

44. Rocks made of loose materials that have been cemented together are:

A. igneous
B. metamorphic
C. sedimentary

45. River valley glaciers produce

A. U-shaped erosion
B. V-shaped erosion
C. S-shaped erosion

46. The life cycle of a river with the most cutting power and erosion is known as which stage?

A. youth stage
B. mature stage
C. old age

**47. The result of radioactive decay**

A. parent element
B. daughter element
C. half-life

**48. The most abundant dry gas found in the atmosphere is**

A. oxygen
B. nitrogen
C. $CO_2$

**49. A natural groundwater outlet through which boiling water and steam explodes into the air is called a _____.**

A. sinkhole
B. artesian system
C. geyser

**50. Which of the following rocks make the best aquifer?**

A. granite
B. basalt
C. sandstone

**51. Sediments that settle out from rivers are called**

A. deposits
B. boulders
C. sandstone

**52. A hole that remains in the ground after a block of glacier ice melts is called a**

A. pothole
B. sinkhole
C. kettle

**53. The first sign that a tsunami is approaching a shore is**

A. a sudden flatten of the waves
B. water moving from the shore
C. large wall of water on horizon

**54. Mountains that have been squeezed into wavelike patterns are called**

A. fold mountains
B. dome mountains
C. fault-block mountains

**55. The largest ocean is the**

A. Atlantic
B. Pacific
C. Indian

**56. The major surface current that flows along the east coast of the United States is known as the**

A. Bermuda Current
B. Mexican Current
C. Gulf Stream

**57. The formation of ocean waves is caused by**

A. Earth's rotation
B. the moon
C. the wind

**58. The most abundant compound found in sea water is**

A. chloride
B. calcium carbonate
C. magnesium chloride

**59.** The distance between two meridians is measured in degrees of

A. longitude
B. latitude
C. magnitude

**60.** A contour line that has tiny comb-like lines along the inner edge indicates a

A. depression
B. mountain
C. valley

**61.** Fossils that are used to date strata are called

A. datum fossils
B. index fossils
C. true fossils

**62.** Which of the following causes the aurora borealis?

A. gases escaping from earth
B. particles from the sun
C. particles from the moon

**63.** The layer of the atmosphere that shields earth from harmful ultraviolet radiation is called

A. ionic layer
B. ozone layer
C. equatorial layer

**64.** The layer of the earth's atmosphere that is closest to the earth's surface is the

A. stratosphere layer
B. thermosphere layer
C. troposphere layer

**65.** The sun transfers its heat to other objects by

A. conduction
B. radiation
C. convection

**66.** As an air mass expands, it becomes

A. cooler
B. warmer
C. denser

**67.** Air moving northward from the horse latitudes produces a belt of winds called the

A. prevailing westerlies
B. north westerlies
C. trade winds

**68.** Which type of cloud always produces precipitation?

A. altostratus
B. cirrostratus
C. nimbostratus

**69.** An air mass that forms over the Gulf of Mexico is called

A. polar
B. maritime
C. continental

**70.** Spring tides will occur when the moon is in its

A. quarter phases
B. full and new phases
C. half phases

**71. Air pressure is measured using a**

A. barometer
B. hydrometer
C. physcrometer

**72. The two most abundant elements found in stars are**

A. hydrogen and calcium
B. hydrogen and helium
C. hydrogen and neon

**73. A comet's tail always points _____ from the sun.**

A. towards
B. perpendicular
C. away

**74. The dark areas observed on the sun are known as**

A. solar flares
B. prominences
C. sun spots

**75. An example of distance in degrees of latitude is**

A. 55° north
B. 93° east
C. 25° west

**76. A scale use to measure the hardness of a mineral is known as the**

A. Bowen's scale
B. Mohs' scale
C. Harding scale

**77. When a gas changes to a liquid this process is known as**

A. evaporation
B. condensation
C. dissolution

**78. A fan-shaped river deposit is better known as a**

A. levee
B. flood plain
C. delta

**79. When heat energy is trapped by the gases in the Earth's atmosphere this process is called**

A. greenhouse effect
B. coriolis effect
C. constant effect

**80. Winds in the Northern Hemisphere are deflected to the**

A. north
B. left
C. right

**81. Water vapor and _____ trap heat in the atmosphere.**

A. carbon dioxide
B. nitrogen
C. sodium nitrate

**82. The frontal system that forms when a cold air mass meets a warm air mass and does not change position is defined as a**

A. occluded front
B. stationary front
C. warm front

EARTH SCIENCE

83. Surface ocean currents are caused by which of the following

A. temperature
B. density changes in water
C. wind

84. The length of time it takes for two waves to pass in a row is called

A. wave length
B. wave period
C. wave crest

85. Circulation of the deep ocean currents is the result of

A. equatorial currents
B. surface currents
C. density currents

86. Chains of undersea mountains associated with the spreading of the seafloor are known as _____.

A. ocean trenches
B. mid ocean ridges
C. seamounts

87. A shallow, calm area of water located between a barrier island and a beach area is defined as a/an _____.

A. atoll
B. coral reef
C. lagoon

88. Closed contour lines noticed on a topographical map indicate which type of information?

A. rivers and lakes
B. hills
C. mountains

89. The heliocentric model was developed by which famous scientist?

A. Kepler
B. Copernicus
C. Newton

90. The phases of the moon are the result of its _____ in relation to the sun.

A. revolution
B. rotation
C. position

91. A telescope that collects light by using a concave mirror and can produce small images is called a _____.

A. radioactive telescope
B. reflecting telescope
C. refracting telescope

92. The measuring unit to measure the distance between stars is called

A. astronomical unit
B. light-year
C. parsec

93. The largest planet found in the solar system is

A. Pluto
B. Jupiter
C. Saturn

94. The famous scientist who discovered the elliptical orbits

A. Kepler
B. Copernicus
C. Galilee

95. The planet with retrograde rotation is

A. Pluto
B. Uranus
C. Venus

96. A star's brightness is referred to as

A. magnitude
B. mass
C. apparent magnitude

97. Clouds of gas and dust where new stars originate are called

A. black holes
B. super novas
C. nebulas

98. The transfer of heat from the earth's surface to the atmosphere is called ____.

A. conduction
B. radiation
C. convection

99. The ozone layer is found in the

A. stratosphere layer
B. mesosphere layer
C. exosphere layer

100. The coldest zone of the atmosphere is found in the

A. thermosphere
B. mesosphere
C. stratosphere

101. Winds in high pressure areas tend to blow

A. clockwise
B. counterclockwise
C. along the center

102. When warm air meets cold air this is defined as a

A. cold front
B. occluded front
C. warm front

103. The fastest velocity of a river is found where?

A. bottom
B. center
C. sides

104. As a glacier melts the sea level tends to:

A. rise
B. sink
C. evaporate

105. The largest groups of minerals found in the earth's crust are

A. silicates
B. carbonates
C. quartz

106. Used to measure the magnitude of an earthquake.

A. Richter scale
B. epicometer
C. seismograph

107. These are types of folds:

A. anticlines and synclines
B. faults and folds
C. fractures and shearings

108. Breaks in rocks which indicate movement are known as

A. fractures
B. folds
C. faults

109. The collision of two continental plates is called a

A. folded mountain range
B. volcanic mountain range
C. block mountain range

110. Plates that move in the same direction are termed

A. divergent faults
B. convergent faults
C. transform faults

111. Studying the positions of layered rock is referred to as

A. relative ages
B. index fossils
C. disconformity

112. The smallest division of geologic time is defined as

A. Periods
B. Eras
C. Epochs

113. The most common fossils of the Paleozoic Era are

A. angiosperms
B. trilobites
C. endotherms

114. Contamination may enter groundwater by

A. air pollution
B. leaking septic tanks
C. photochemical processes

115. Which is a form of precipitation?

A. snow
B. frost
C. fog

116. A dead star is called a _____.

A. White Dwarf
B. Super Giant
C. Black Dwarf

**117.** Roughly ninety percent of all geologic time is said to be _____.

A. Paleozoic
B. Pre-Cambrian
C. Mesozoic

**118.** The massive change in biological conditions that marked the beginning of life forms on earth is known as _____.

A. Oxygen Revolution
B. Carbon Revolution
C. Trilobite Revolution

**119.** Water is a truly unique material. It has the property of _____.

A. Adhesion
B. Cohesion
C. Both

**120.** The following is not a form of satellite used to track weather:

A. NEXRAD
B. Geostationary
C. Polar Orbitting

**121.** Over the course of our planet's history Earth has had _____ atmosphere(s).

A. one
B. two
C. three

**122.** Which is not a principle law of geology?

A. Cross Cutting
B. Faulting
C. Super position

**123.** The red beds are important because they indicate the presence of _____ in the geologic record.

A. Carbon
B. Ammonia
C. Oxygen

**124.** Tornadoes are most likely to occur in what season?

Spring
Summer
C. Autumn

**125.** Which scale is used to measure hurricanes?

A. Fujita Scale
B. Saffir-Simpson Scale
C. Richter Scale

## Answer Key

| | | |
|---|---|---|
| 1. A | 45. A | 89. B |
| 2. B | 46. A | 90. C |
| 3. C | 47. B | 91. B |
| 4. B | 48. B | 92. C |
| 5. A | 49. C | 93. B |
| 6. C | 50. C | 94. A |
| 7. C | 51. A | 95. C |
| 8. B | 52. C | 96. A |
| 9. B | 53. B | 97. C |
| 10. C | 54. A | 98. A |
| 11. A | 55. B | 99. A |
| 12. B | 56. C | 100. B |
| 13. A | 57. C | 101. A |
| 14. C | 58. A | 102. C |
| 15. A | 59. A | 103. B |
| 16. A | 60. A | 104. A |
| 17. A | 61. B | 105. A |
| 18. C | 62. B | 106. C |
| 19. B | 63. B | 107. A |
| 20. B | 64. C | 108. C |
| 21. C | 65. B | 109. A |
| 22. C | 66. B | 110. C |
| 23. A | 67. A | 111. A |
| 24. C | 68. C | 112. C |
| 25. A | 69. B | 113. B |
| 26. A | 70. B | 114. B |
| 27. A | 71. A | 115. A |
| 28. C | 72. B | 116. C |
| 29. C | 73. C | 117. B |
| 30. A | 74. C | 118. A |
| 31. C | 75. A | 119. C |
| 32. B | 76. B | 120. A |
| 33. C | 77. B | 121. C |
| 34. B | 78. C | 122. B |
| 35. C | 79. A | 123. C |
| 36. C | 80. C | 124. A |
| 37. A | 81. A | 125. B |
| 38. C | 82. B | |
| 39. B | 83. C | |
| 40. B | 84. B | |
| 41. C | 85. C | |
| 42. A | 86. B | |
| 43. B | 87. C | |
| 44. C | 88. B | |

# TEACHER CERTIFICATION STUDY GUIDE

## Rationales with Sample Questions

1. Which layer of the atmosphere would you expect most weather to occur?
A. Troposphere
The troposphere is the lowest portion of the Earth's atmosphere. It contains the highest amount of water and aerosol. Because it touches the Earth's surface features, friction builds. For all of these reasons, weather is most likely to occur in the Troposphere.

2. What percentage of earth's surface is covered by water?
B. 71%
The earth's surface is nearly ¾ covered with water. The Pacific Ocean is the largest body of moving water. Of course there are other oceans, lakes, rivers, and glaciers as well.

3. Which layer of the earth's atmosphere contains the Ozone layer?
C. Stratosphere
The stratosphere is located above the troposphere and below the mesosphere. It has layers striated by temperature. The warmest portion, the ozone layer, is warm because it absorbs solar ultraviolet radiation.

4. Copernicus developed a theory that is known as
B. Heliocentric
Copernicus' theory stated that the planets revolved around the sun (helios), as opposed to prior belief that the planets revolved around the Earth (geocentric).

5. The boundary that separates the crust from the mantle is known as
A. Moho
The Mohorovicic Discontinuity separates oceanic and/or continental crust from the Earth's mantle.

6. The product of intrusive activities would result in forming a
C. dike
A dike is formed when upwelling magma cools and solidifies beneath the surface, an intrusive activity.

7. A star's light and heat are produced by
C. nuclear fusion
Nuclear fusion is the process in which hydrogen atoms fuse together to form helium atoms, releasing massive amounts of energy during the fusion. It's the fusion of atoms, not combustion, which causes the star to shine.

8. The center of an atom is called
B. nucleus
The center of the atom is the nucleus. The nucleus of the atom is composed of nucleons, which when electrically charged are protons and when electrically neutral are neutrons. However, the electrons swirl around the nucleus in a large region called the Electron Cloud.

9. An important food source for animals in a water biome
B. plankton
Drifting organisms that inhabit the water column are called plankton. They may be phytoplankton or zooplankton. Phytoplankton are autotrophs and form the base of the aquatic food chain.

10. The smallest piece of an element is called a/an
C. atom
An atom is the smallest particle of the element that has the properties of that element. All of the atoms of a particular element are the same. The atoms of each element are different from the atoms of the other elements.

11. An instrument that measures relative humidity is known as
A. psychrometer
A psychrometer measures relative humidity. The other choices, anemometer and barometer, measure wind speed and atmospheric pressure, respectfully.

12. Removing salts from ocean water by heating is called
B. distillation
In the process of distilling ocean water the saline water is heated, producing water vapor that is in turn condensed, forming fresh water. The salt is left behind as waste but the water is used in many areas for drinking supply.

13. When molecules in the air cool and combine to form rain, _____ has occurred.
A. condensation
Condensation is the change in matter from a denser phase, such as a gas (or vapor) to a liquid. Condensation commonly occurs when a vapor is cooled to a liquid.

14. Which instrument measures wind direction?
C. wind vane
Of the choices given, an anemometer measures wind speed (velocity), a barometer measures atmospheric pressure and a wind vane indicates wind direction.

15. The most important cause of erosion is
A. water
Erosion is most often caused by water. This can be acid rain eroding rocks, rivers eroding riverbeds, oceans eroding beaches and cliffs, etc. In addition, wind is another source of erosion.

16. An anemometer measures
A. wind velocity
Of the choices given, an anemometer measures wind speed (velocity), temperature would be measured by a thermometer, and relative humidity is measured with a psychrometer.

17. The boundary that develops when a cold air mass meets a warm air mass
A. cold front
Fronts are always labeled according to the approaching air mass. Therefore, a cold air mass meeting and displacing a warm air mass would be called a cold front.

18. When the sun, moon and earth are aligned in a straight line what type of tides are produced?
C. spring tides
Spring tides are produced when the Earth, Sun, and Moon are in a line. Therefore, spring tides occur during the full moon and the new moon. Neap tides occur during quarter moons. They occur when the gravitational forces of the Moon and the Sun are perpendicular to one another (with respect to the Earth).

19. North of the equator, currents move in which direction?
B. clockwise
North of the equator, currents move clockwise. South of the equator, currents move counter clockwise.

20. Rocks that serve as aquifers are
B. permeable
Aquifers are underground areas of water-bearing permeable rock from which groundwater can be collected.

21. Volcanoes with violent eruptions are known as
C. cinder volcanoes
Cinder volcanoes are some of the most violent volcanoes because of the immense pressure of gas built up within the neck of the volcanic tube. When it overcomes the resistance offered by the surrounding rock, it rips off the top of the cone. A huge mass of liquid magma and Pyroclastic Rock are flung outward in a violent explosion.

22. The Richter scale measures
C. magnitude
The richter scale is used to measure the magnitude of earthquakes. Focus and compressions refer to areas of activity, but are not examples of a scale for measuring.

23. The Earth's outer core is probably
A. liquid
The earth's inner core is mathematically hypothesized to be a solid iron and nickel core. The outer core, surrounding the inner core, is so hot that it is believed to be molten iron (liquid state). Combined, they are responsible for Earth's magnetism.

24. What are the two most abundant elements found in the earth's crust?
C. oxygen and silicon
Earth's crust is composed of 47% oxygen and 28% silicon.

25. The San Andreas Fault is classified as a
A. transform fault
The San Andreas fault is considered a transform fault because sections of the earth's crust (the Pacific and North American Plates) slide side-by-side past each other.

26. Batholiths are the largest structures of which type of rock activity?
A. intrusive rock
Batholiths are large portions of igneous intrusive rock deep within the Earth's crust that form from cooled magma.

27. The main agents of chemical weathering are
A. water, oxygen, $CO_2$
Water is the greatest factor in chemical weathering. Glaciers erode entire valleys. Rainfall pounds way at topographic surface features. Rivers erode riverbeds and river edges. Oceans erode shorelines and cliffs. Oxygen is also a factor in weathering. Air movement can have erosional factors. Most importantly, wind can transport material to other areas, having both erosional and depositional results. Carbon dioxide combines with water to produce carbonic acid, which erodes rock structures and some of our man made monuments.

28. Soil classified as porous is called
C. sandy soil
Sandy soil has a high sand content. The sand molecules have many spaces in-between, making the soil porous. This soil does not hold water well.

29. Intrusive igneous rock forms
C. large crystals
Intrusive igneous rock forms large crystals. This rock is formed from magma that cools and solidifies within the earth. Because it is surrounded by pre-existing rock, the magma cools slowly, and the rocks are coarse grained. The crystals are usually large enough to be seen by the unaided eye.

30. Which types of rocks are rich sources of fossil remains'?
A. sedimentary rock
Sedimentary rock has the most abundant fossil collection. This is because, over time the layers of sand and mud at the bottom of lakes & oceans turned into rocks due to compression. Plants and animals that died and fell to the bottom were part of the compressional process by which the many layers were eventually turned into stone, encapsulating a fossil.

31. The best preserved animal remains have been discovered in
C. tar pits
Tar pits provide a wealth of information when it comes to fossils. Tar pits are oozing areas of asphalt, which were so sticky as to trap animals. These animals, without a way out, would die of starvation or be preyed upon. Their bones would remain in the tar pits, and be covered by the continued oozing of asphalt. Because the asphalt deposits were continuously added, the bones were not exposed to much weathering, and we have found some of the most complete and unchanged fossils from these areas, including mammoths and saber toothed cats.

32. The Mid-Atlantic is a major area of which type plate movement?
B. divergent plate
The Mid- Atlantic is home to a submerged mountain range, which extends from the Arctic Ocean to beyond the southern tip of Africa. The divergent plate action results in sea floor spreading at a rate of about 2.5 centimeters per year (cm/yr), or 25 km in a million years, creating the vast ocean we recognize today.

33. When lava cools quickly on the earth's surface the newly formed rock is called
C. extrusive
Rock formed by the cooling of magma on the earth's surface is known as extrusive, as opposed to intrusive, which is formed by the cooling of magma below the Earth's surface.

34. When a dyke forms with magma flowing in a tub-like structure this is known to be a/an
B. intrusive activity
Dykes are thin, vertical veins of igneous rock. They form within fractures in the earth's crust. Intrusive activity forces magma into underground areas, which can seep into these existing fractures forming a dyke.

35. A caldera is formed when a large depression collapses. This is the result of a
C. volcanic eruption
A caldera is the collapse of land following a volcanic eruption. Once the underground store of magma and gas has been released in a volcanic explosion, there is not enough support, causing the ground to collapse. A caldera is sometimes confused with the area from which magma and gases are emitted (a crater).

36. Trenches observed on the sea floor are the results of
C. subduction
Trenches are created where two plates collide (converge). Plate collision causes denser oceanic crust to sink or slip beneath lighter continental crust. It is subducted and melted into the asthenosphere, producing a deep trench on the ocean floor parallel to the plate boundary.

37. Alfred Wegener's hypothesis of continental drift was not supported until scientists began studying
A. sea floor
Wegener's hypothesis of continental drift was supported by studies of the sea floor. In comparison to continental rock materials, the youngest rock is found on the ocean floor, consistent with the tectonic theory of cyclic spreading and subduction. Overall, oceanic material is roughly 200 million years old, while most continental material is significantly older, with age measured in billions of years.

38. These massive waves are caused by the displacement of ocean water, and are often the result of underwater earthquakes.
C. tsunamis
Earthquakes can trigger an underwater landslide or cause sea floor displacements that in turn, generate deep, omni-directional waves. Far out to sea these waves may be hardly noticeable. However, as they near the shoreline, the shallowing of the sea floor forces the waves upward in a "springing" type of motion.

39. Sea floor spreading occurs when the earth's crust is stretched and pulled apart in a process called
B. rifting
A rift is a place where the Earth's crust and lithosphere are being pulled apart. In rifts, no crust or lithosphere is produced. If rifting continues, eventually a mid-ocean ridge may form.

40. The layer of the atmosphere that is in a plasma state and aids in communication
B. Ionosphere
The Ionosphere is an area of free ions: positively charged ions, produced as a result of solar radiation striking the atmosphere. It is known for its production of aurora borealis and its benefits to radio transmission

41. A stream erodes bedrock by grinding sand and rock fragments against each other. This process is defined as
C. abrasion
Abrasion is the key form of mechanical weathering. It is a sandblasting effect caused by particles of sand or sediment. Abrasive agents include wind blown sand, water movement, and the materials in landslides bashing into each other.

42. Rocks formed from magma are
A. igneous
Igneous rocks are rocks that have formed from cooled magma. They are further classified as extrusive or intrusive according to location.

43. Rocks formed by the intense heating or compression of pre- existing rocks are classified as
B. metamorphic
Metamorphism is the process of changing a pre-existing rock into a new rock by heat and or pressure. Metamorphism is similar to that of putting a clay pot into a kiln. The clay doesn't melt, but a solid-state chemical reaction occurs that causes a change. The chemical bonds of adjoining atoms breakdown and allow the atoms to rearrange themselves, producing a substance with new properties.

44. Rocks made of loose materials that have been cemented together
C. sedimentary
Sediments are broken up rock material. Sand on a beach or pebbles in a mountain stream are typical examples. Sedimentary rocks are named for their source; they are rocks that form from sediments that lithify to become solid rock. Sedimentary rock is especially important for the finding of fossils.

45. River valley glaciers produce
A. U-shaped erosion
River valleys are typically V- shaped. The velocity and cutting power of a river is greatest at its center. However, glaciers broaden the area. Upon its retreat, a glacier typically leaves a U- shaped eroded valley.

46. The life cycle of a river with the most cutting power and erosion is known as which stage?
A. youth stage
Young streams have straight paths, no flood plain, a "V" shaped cutting profile, and high velocity with generally clear water and low suspended load. Old streams have lots of meanders, large flood plain, flat profile, low velocity, with murky, "muddy" waters because of a high-suspended load.

# TEACHER CERTIFICATION STUDY GUIDE

47. The result of radioactive decay
B. daughter element
The radioactive decay causes the (mother) element to change into an (daughter) element. The Mother-Daughter relationship of produced nuclides during the series of isotope decay is the basis for radiometric dating. Although many isotopes are used in radiometric dating, the most widely known method is referred to as Carbon-14 dating. Knowing the half-life (how long it takes for half of the material to decay) is the key factor in the radiometric dating process.

48. The most abundant dry gas found in the atmosphere is
B. Nitrogen
The atmosphere is composed of 78% Nitrogen, 21% Oxygen, and 1% other gasses.

49. A natural groundwater outlet through which boiling water and steam explodes into the air is called a
C. geyser
A geyser is a thermal spring that erupts. The processes behind the eruption are very similar to those involved in boiling water in a teakettle. A constriction forms in the connected chambers of a spring. The water heats under pressure, turns to steam, and erupts with great force past the constriction. The ejected steam condenses and returns to a liquid state. The water draws back into its chambers and the process begins again. Since it takes awhile for the water to drain back and reheat, geysers often erupt on a determinable schedule.

50. Which of the following rocks make the best aquifer?
C. sandstone
Sandstone makes the best aquifer because of its porosity. It has larger pores than granite or basalt, and is also likely to fracture in a way that is conducive to water movement and collection.

51. Sediments that settle out from rivers are called
A. deposits
Deposits are pieces of matter that settle out of the water and fall to the bottom, or are washed into a collection area, such as a delta. This can be terrestrial matter, biological matter, salts, or larger pebbles and rocks.

52. A hole that remains in the ground after a block of glacier ice melts is called a
C. kettle
As the outwash moves sediment alongside and in the path of a receding glacier, blocks of ice can be buried beneath the sediment. After years of erosion these blocks are uncovered and melt, leaving a shallow depression behind. When these depressions fill, they are known as Kettles, and become scenic lakes.

53. The first sign that a tsunami is approaching a shore is
B. water moving from the shore
The first sign that a tsunami is approaching is usually the retreat of water from the shoreline. When the water returns, it comes fast and washes well past its normal level in both distance and depth, destroying coastal areas and causing many losses.

54. Mountains that have been squeezed into wavelike patterns are called
A. fold mountains
During mountain building or compressional stress, rocks may deform to produce folds. Generally, a series is produced. The up-folds are called anticlines and the down-folds are known as synclines.

55. The largest ocean is the
B. Pacific
The four major oceans (listed in decreasing size) are the Pacific, Atlantic, Indian and Arctic.

56. The major surface current that flows along the east coast of the United States is known as the
C. Gulf Stream
The Gulf Stream begins in the Caribbean and ends in the northern North Atlantic. It is powerful enough to be seen from outer space and is one of the world's most studied current systems. It acts as the east coast boundary current plays an important role in the transfer of heat and salt to the poles.

57. The formation of ocean waves is caused by
C. the wind
Wind is the primary factor in the production of ocean waves. It is the energy and friction of wind action that transfers to the water to create waves.

58. The most abundant compound found in sea water is
A. chloride
Chloride is the compound found most often in sea water. Other compounds commonly found include sodium carbonate, magnesium and potassium compounds, sulfite, bromide, and silicate. NaCl is what we commonly refer to as sea salt. Of the two components, chloride is more readily available in the sea.

59. The distance between two meridians is measured in degrees of
A. longitude
Longitude describes the location of a place on Earth east or west of a line called the Prime Meridian. Longitude is given in degrees ranging from 0° at the Prime Meridian to 180° east or west

60. A contour line that has tiny comb-like lines along the inner edge indicates a
A. depression
Contour lines are shown as closed circles in elevated areas and as lines with miniature perpendicular lined edges where depressions exist. These little lines are called hachure marks.

61. Fossils that are used to date strata are called
B. index fossils
Index fossils are fossils of organisms that were known to be abundant at specific times in Earth's history. Presence of such fossils gives one an idea of what age the surrounding material came from.

62. Which of the following causes the aurora borealis?
A. particles from the sun
Aurora Borealis is a phenomenon caused by particles escaping from the sun. The particles escaping from the sun include a mixture of gases, electrons and protons, and are sent out at a force that scientists call solar wind. Together, we have the Earth's magnetosphere and the solar wind squeezing the magnetosphere and charged particles everywhere in the field. When conditions are right, the build-up of pressure from the solar wind creates an electric voltage that pushes electrons into the ionosphere. Here they collide with gas atoms, causing them to release both light and more electrons.

63. The layer of the atmosphere that shields earth from harmful ultraviolet radiation is called
B. ozone layer
The ozone layer is the part of the Earth's atmosphere that contains high concentrations of ozone ($O_3$). It is located in the stratosphere and absorbs UV radiation emitted from the sun, making life possible on Earth.

64. The layer of the earth's atmosphere that is closest to the earth's surface is the
C. troposphere layer
The troposphere is the layer of Earth's atmosphere that is the lowest (closest to the surface). It is the densest because it contains almost all the water vapor and aerosol found in the atmosphere. It is easy to conclude, then, that most weather phenomena occur here.

65. The sun transfers its heat to other objects by
B. radiation
Radiation is the process by which energy is transferred in the form of waves or particles. The Sun emits ultraviolet radiation in UVA, UVB, and UVC forms, but because of the ozone layer, most of the ultraviolet radiation that reaches the Earth's surface is UVA.

66. As an air mass expands it becomes
B. warmer
Air expends as heat is applied according to the laws of gasses.

67. Air moving northward from the horse latitudes produces a belt of winds called the
A. prevailing westerlies
The prevailing westerlies are the winds found in the middle latitudes between 30 and 60 degrees latitude. They blow from the high pressure area in the horse latitudes towards the poles.

68. Which type of cloud always produces precipitation?
C. nimbostratus
Nimbostratus clouds are seen as a thick, uniform, gray layer from which precipitation (significant rain or snow) is falling. Of the other choices offered, altostratus clouds appear as uniform white or bluish-gray layers that partially or totally obscure the sky, and cirrostratus are like a thin, nearly transparent, veil or sheet that partially or totally covers the sky. Only nimbostratus guarantees precipitation.

69. An air mass that forms over the Gulf of Mexico is called
B. maritime
Maritime air masses are moist, containing considerable amounts of water vapor, which is ultimately condensed and released as rain or snow. Maritime tropical air originates near the Gulf of Mexico and travels north-east across the warm Atlantic to affect western Europe, as well as north-west across the United States.

70. Spring tides will occur when the moon is in its
B. full and new phases
Spring tides are produced when the Earth, Sun, and Moon are in a line. Therefore, spring tides occur during the full moon and the new moon. Neap tides occur during quarter moons. They occur when the gravitational forces of the Moon and the Sun are perpendicular to one another (with respect to the Earth).

71. Air pressure is measured using a
A. barometer
A psychrometer measures relative humidity. A barometer measures atmospheric pressure. A hydrometer is used to measure the specific gravity of a liquid.

72. The two most abundant elements found in stars are
B. hydrogen and helium
Hydrogen and helium are the only elements that occur naturally in our universe. It makes sense, then, that they are present in all areas, including stars.

73. A comet's tail always points _____ from the sun.
C. away
A comet's tail always points away from the sun. The sun's radiation is burning up the ice that makes the comet, and since it is projecting the material outward, the tail seems to be pointing away from the sun. Notice that this question does not use a specific direction (north, south, east, west) because comets move and are subject to the viewer's location and perception.

74. The dark areas observed on the sun are known as
C. sun spots
Larger dark spots called Sunspots appear regularly on the Sun's surface. These spots vary in size from small to 150,000 kilometers in diameter and may last from hours to months. The sunspots also cause solar flares that can accelerate to velocities of 900 km/hr, sending shock waves through the solar atmosphere.

75. An example of distance in degrees of latitude is
A. 55° north
Latitude is measured in degrees away from the equator. The equator marks 0°, and parallel lines moving around the globe are quantified in degrees north or south.

76. A scale use to measure the hardness of a mineral is known as the
B. Moh's scale
The Moh's scale of hardness measures the scratch resistance of minerals. The hardest material is diamond, and the frailest is talc. This means that diamond can scratch any surface, which is not true of less hard materials, such as talc.

77. When a gas changes to a liquid this process is known as
B. condensation
Condensation is the change in matter to a denser phase, such as a gas (or vapor) to a liquid. Condensation can occur when a vapor is cooled to a liquid or when a vapor is compressed.

78. A fan-shaped river deposit is better known as a
C. delta
Flowing water carries the material to the ocean where one of two things happen, the material is deposited on the offshore continental shelf or is carried back inland to the inlets and bays. Over time, the sediment thickly accumulates and may form typical coastal features such as sand bars and deltas.

EARTH SCIENCE

79. When heat energy is trapped by the gases in the Earth's atmosphere this process is called
A. greenhouse effect
When greenhouse gases and heat build up, the Earth's surface and atmospheric temperature rises. The current and controversial hypothesis contends that if we cut the amount of rising $CO_2$ in the atmosphere, then things will cool down.

80. Winds in the Northern Hemisphere are deflected to the
C. right
The Earth is spinning on its rotational axis. Spin is greatest near the equator and least at the poles. The different velocities associated with the spin give rise to an effect on the air known as the Coriolis Force. The idea is that the result of the Coriolis effect is that winds in the north are deflected to the right, and winds in the south are deflected to the west.

81. Water vapor and _____ trap heat in the atmosphere.
A. carbon dioxide
Water vapor and carbon dioxide are both considered greenhouse gases because they can trap heat in the atmosphere. Other sources of greenhouse gasses include rice paddies and ruminant animals, which produce Methane.

82. The frontal system that forms when a cold air mass meets a warm air mass and does not change position is defined as a
B. stationary front
Fronts are the boundaries where one air mass meets another. A stationary front is a boundary between two air masses when neither is strong enough to displace the other.

83. Surface ocean currents are caused by which of the following
C. wind
A current is a large mass of continuously moving oceanic water. Surface ocean currents are mainly wind-driven and occur in all of the world's oceans (example: the Gulf Stream). This is in contrast to deep ocean currents which are driven by changes in density.

84. The length of time it takes for two waves to pass in a row is called
B. wave period
The wave period is the time required for two successive waves to pass. Wave crest is the tallest part of the wave. Wave length is measured from the crest of one wave to the crest of the next.

**EARTH SCIENCE**

85. Circulation of the deep ocean currents is the result of
C. density currents
Unlike surface currents, deep ocean currents are driven by changes in density. These density differences may be caused by changes in salinity (halocline) or temperature (thermocline). Colder water sinks below warmer waters, causing a river (current) flowing below the warmer waters.

86. Chains of undersea mountains associated with the spreading of the seafloor are known as
B. mid ocean ridges
Mid ocean ranges are underwater mountains formed by plate tectonics. The underwater mountains are all connected, making a single mid-oceanic ridge system that is the longest mountain range in the world. The ridges are active sites with new magma constantly emerging onto the ocean floor and into the crust, resulting in sea floor spreading.

87. A shallow, calm area of water located between a barrier island and a beach area is defined as a/an
C. lagoon
A lagoon is known for its quiet movement of water. A lagoon is a body of shallow salt or brackish water separated from the sea by a shallow or exposed sandbank, coral reef, etc. Non-reef lagoon barriers are formed by wave-action or longshore currents depositing sediments. Because of their gentle atmosphere and brackish water, they are often nurseries for many baby fish and aquatic animals.

88. Closed contour lines noticed on a topographical map indicate which type of information?
B. hills
The rules of contouring dictate that contour lines are closed around hills, basins, or depressions. Because we know that depressions are shown using hachure marks, a closed contour line without such marks represents a hill.

89. The heliocentric model was developed by which famous scientist?
B. Copernicus
Copernicus is recognized for his heliocentric theory. The heliocentric theory postulates that the heavenly bodies rotate around the sun. Prior to his assertions, people believed in the geocentric model that held that all bodies rotated around the Earth. The geocentric model was supported by the church, so Copernicus' ideas were highly controversial.

90. The phases of the moon are the result of its _____ in relation to the sun.
C. position
The moon is visible in varying amounts during its orbit around the earth. One half of the moon's surface is always illuminated by the Sun (appears bright), but the amount observed can vary from full moon to none.

# TEACHER CERTIFICATION STUDY GUIDE

91. A telescope that collects light by using a concave mirror and can produce small images is called a
B. reflecting telescope
Reflecting telescopes are commonly used in laboratory settings. Images are produced via the reflection of waves off of a concave mirror. The larger the image produced the more likely it is to be imperfect.

92. The measuring unit to measure the distance between stars is called
C. parsec
Parsecs are the units used to describe the distance between stars. Astronomical units (AU) are used to describe the distances between celestial objects (example The Earth is 1.00 ± 0.02 AU from the Sun ). Light years are a unit of length measuring the distance light travels in a vacuum in one year.

93. The largest planet found in the solar system is
B. Jupiter
The planets (in decreasing size) are Jupiter, Saturn (body- not inclusive of rings), Uranus, Neptune, Earth, Venus, Mars, Mercury (Pluto was thought to be the smallest planet, but is no longer classified as a planet).

94. The famous scientist who discovered the elliptical orbits
A. Kepler
The significance of Kepler's Laws is that it overthrew the ancient concept of uniform circular motion, which was a major support for the geocentric arguments. Although Kepler postulated three laws of planetary motion, he was never able to explain *why* the planets move along their elliptical orbits, only that they did.

95. The planet with retrograde rotation is
C. Venus
Venus has an axial tilt of only 3° and a very slow rotation. It spins in the direction opposite of its counterparts (who spin in the same direction as the Sun). Uranus is also tilted and orbits on its side. However, this is thought to be the consequence of an impact that left the previously prograde rotating planet tilted in such a manner.

96. A star's brightness is referred to as
A. magnitude
Magnitude is a measure of a star's brightness. The brighter the object appears, the lower the number value of its magnitude. The apparent magnitude is how bright an observer perceives the object to be. Mass has to do with how much matter can be measured, not brightness.

97. Clouds of gas and dust where new stars originate are called
C. nebulae
Nebulae are where new stars are born. They are large areas of gasses and dust. When the conditions are right, particles combine to form stars.

EARTH SCIENCE

98. The transfer of heat from the earth's surface to the atmosphere is called
A. conduction
Radiation is the process of warming through rays or waves of energy, such as the Sun warms earth. The Earth returns heat to the atmosphere through conduction. This is the transfer of heat through matter, such that areas of greater heat move to areas of less heat in an attempt to balance temperature.

99. The ozone layer is found in the
A. stratosphere
The stratosphere is home to the ozone layer, which protects Earth from harmful UV radiation.

100. The coldest zone of the atmosphere is found in the
B. mesosphere
The mesosphere is the coldest layer of the atmosphere, with temperatures as low as -100°Celsius. Within this layer, temperature decreases with increasing altitude.

101. Winds in high pressure areas tend to blow
A. clockwise
High pressure systems are known for winds that flow clockwise and fair weather. Low pressure systems are accompanied by clouds and precipitation and winds flow counterclockwise.

102. When warm air meets cold air this is defined as a
C. warm front
When a warm air mass meets and displaces a cold air mass, the front is called a warm front.

103. The fastest velocity of a river is found where?
B. center
Mountain streams have little fining (sorting the material by size) due to their higher velocity, and low land streams are muddy because the velocity is less and erosion occurs on the bed and sides of the stream. Once a stream is at or close to base level, equilibrium is achieved between deposition and erosion. Erosion and deposition are controlled by the velocity of the stream. As the stream approaches base level, more of its energy is in a side-to-side cutting (meanders) than in down-cutting.

104. As a glacier melts the sea level tends to
A. rise
As a glacier melts, its water is distributed into nearby bodies of water, causing the sea level to rise.

105. The largest groups of minerals found in the earth's crust are
A. silicates
Silicates are the most abundant group of minerals found in the Earth's crust. The two most abundant elements in the earth's crust are Oxygen (46.6%) and Silicon (27.7%). These combine together to form silicates, which some scientists believe make up as much as 90% of the Earth's crust.

106. Used to measure the magnitude of an earthquake.
C. seismograph
A seismograph is a machine used to measure the magnitude of an earthquake. As the Earth's materials move, the weight also moves and sends an electronic signal to a recording device called a seismograph. Movements are displayed as a series of lines on a recording chart called a Seismogram, reflecting the seismic energy detected at a particular location.

107. These are types of folds:
A. anticlines and synclines
Folded mountains are composed of up and down folds. The up-folds are called anticlines and the down-folds are known as synclines.

108. Breaks in rocks which indicate movement are known as
C. faults
Faults are rock fractures that indicate relative movement. Fractures are also breaks in the rock, but they show no evidence of movement. Folds are created from compression and are a forming of tectonic building.

109. The collision of two continental plates is called a
A. folded mountain range
The collision of two continental plates results in a folded mountain range. Two continental plates pushing against each other but not subducting, will cause the material to buckle, sometimes repeatedly, giving these mountains their characteristic ribbon appearance.

110. Plates that move in the same direction are termed
C. transform faults
Transform faults are areas where two plates move in the same direction. They are parallel, and do not collide, but may result in earthquakes if areas of the plates stick or have excessive pressure in sliding past each other.

111. Studying the positions of layered rock is referred to as
A. relative ages
The Earth's materials-rocks, soils, and sediments-are piled upon each other in layers called strata. Understanding the relative orientation and arrangement of the strata provides important information about the Earth's history and the ongoing sequence of events and processes that helped shape that history.

# TEACHER CERTIFICATION STUDY GUIDE

112. The smallest division of geologic time is defined as
C. epochs
Geologic time is divided into eons, eras, periods, and epochs (listed here in decreasing order of size).

113. The most common fossils of the Paleozoic Era are
B. trilobites
Trilobites flourished in the Paleozoic era. There were over 600 genera and 1000's of species. Trilobites were bottom dwellers and scavengers found in shallow to deep water. For an extremely long period of time, Trilobites were the dominant multi-cellular life form on the planet. Trilobites are very good guide fossils because they were extremely abundant and existed throughout the entire Paleozoic period. Their development underwent distinctive changes, and these differences are useful in subdividing the time period.

114. Contamination may enter groundwater by
B. leaking septic tanks
Leaking septic tanks allow contamination to slowly seep into the ground, where it is absorbed into the water table and infects the groundwater.

115. Which is a form of precipitation?
A. snow
Snow is a form of precipitation. Precipitation is the product of the condensation of atmospheric water vapor that falls to the Earth's surface. It occurs when the atmosphere becomes saturated with water vapor and the water condenses and falls out of solution. Frost and fog do not qualify as precipitates.

116. A dead star is called a _____.
C. Black Dwarf

The final phase of a lower main sequence star's life cycle can take two paths: most main sequence white dwarfs after a few billion years completely burn out to become what is called a black dwarf: a cold, dead star. Alternatively, if a White Dwarf is part of a Binary Star: two suns in the same solar system, instead of slowly cooling to become a Black Dwarf, it may capture hydrogen from its companion star.

117. Roughly ninety percent of all geologic time is said to be _____.
B. Pre-Cambrian

Pre-Cambrian Time: Comprised of the Hadean, Archean, and Proterozoic Eons, 87% of all geologic time is considered Pre-Cambrian.

EARTH SCIENCE

118. The massive change in biological conditions that marked the beginning of life forms on earth is known as _____.
A. Oxygen Revolution
Between 4.6 and 3.6 billion years ago, we transition from an uninhabitable Earth, to the appearance of simple, single-celled bacteria. Around 2.5 billion years ago, the bacteria developed the ability of photosynthesis. This process released oxygen as a by-product and there was a massive release of oxygen as the bacteria multiplied. This massive release is called the Oxygen Revolution and it concurrently marks the beginning of the Proterozoic Eon.

119. Water is a truly unique material. It has the property of _____.
C. Both
A unique property of water is that water likes itself; it has a natural tendency to stick to itself. This property is based upon the polar nature of the water molecule. It attracts other water molecules. When the molecules stick together, they are attached through Hydrogen Bonds, giving the molecule a property called cohesion. Cohesion gives water an unusually strong surface tension, and its capillary action makes the water spread. When the water spreads, adhesion, the tendency of water to stick to other materials, allows water to adhere to solids, making them wet.

120. The following is not a form of satellite used to track weather:
A. NEXRAD
While all of these instruments are used to track weather, the NEXRAD Radar, Next Generation Doppler Radar, is not a satellite. It emits beams of energy that are reflected by the water droplets in the atmosphere. This type of radar is very useful for tracking and predicting rain and less useful for snow or sleet. Geostationary satellites move with the Earth's rotation. Since they always look at the same point, this allows for a view showing changes over periods of time. Polar Orbiting satellites follow an orbit from pole to pole. The Earth rotates underneath the satellite and gives a view of different areas. In effect, it produces slices of the Earth.

121. Over the course of our planet's history Earth has had _____ atmosphere(s).
C. three
Earth's initial atmosphere was composed of primarily hydrogen and smaller amounts of helium. However, most of the hydrogen and helium escaped into space very shortly after the earth was formed, approximately 4.6 billion years ago. A second atmosphere formed during the first 500 million years of Earth's history, as the gasses trapped within the planet were out-gassed during volcanic eruptions. This atmosphere was composed of carbon dioxide ($CO_2$), Nitrogen (N), and water vapor ($H_2O$), with smaller amounts of methane ($CH_4$), ammonia ($NH_3$), hydrogen (H), and carbon monoxide (CO). However, only trace quantities of oxygen were present. At around 3.5 billion years, Earth's third atmosphere began to form as the first life forms- simple, unicellular bacteria- appeared.

122. Which is not a principle law of geology?
B. faulting

The principle laws of geology are:
- Principle of Uniformitarianism: Processes that are happening today also happened in the past.
- Principle of Cross-Cutting Relations: A rock is younger than any rock it cuts across.
- Principle of Original Horizontality: Rock units are originally laid down flat. Something happened to cause them to change orientation.
- Principle of Super Position: The rock on the bottom is older than the rock on top.
- Principle of Biologic Succession: Fossils correspond to particular periods of time.

123. The red beds are important because they indicate the presence of _____ in the geologic record.
C. Oxygen

Formation of Red Beds: The Animike Group- banded iron formations- form. These Red Beds are important because they herald the appearance of significant amounts of oxygen on the Earth. The red color is produced by rust. The rust indicates the presence of oxygen acting upon the ferrous material present in the ocean, and eventually, on the land. The presence of significant amounts of oxygen allows ozone to form, which in turn, screens out the harmful ultra-violet (UV) rays. This makes life possible outside of the protective confines of the ocean.

124. Tornadoes are most likely to occur in what season?
A. Spring

Tornado: an area of extreme low pressure, with rapidly rotating winds beneath a cumulonimbus cloud. Tornadoes are normally spawned from a Super Cell Thunderstorm. They can occur when very cold air and very warm air meet, usually in the Spring. Tornadoes represent the lowest pressure points on the Earth and move across the landscape at an average speed of 30 mph.

125. Which scale is used to measure hurricanes?

B. Saffir-Simpson Scale

The Fujita Scale is used to measure the intensity and damage associated with tornadoes. The Saffir-Simpson Scale is used to classify hurricanes into five categories, with increasing numbers corresponding to lower central pressures, greater wind speeds, and large storm surges. Richter Scale: the primary scale used by seismologists to measure the magnitude of the energy released in an earthquake.

**XAMonline, INC. 21 Orient Ave. Melrose, MA 02176**

Toll Free number 800-509-4128

*TO ORDER Fax 781-662-9268 OR www.XAMonline.com*

MASSACHUSETTS TEST FOR EDUCATOR LICENTURE - MTEL - 2007

PO#   Store/School:

Address 1:

Address 2 (Ship to other):
City, State Zip

**Credit card number** ___-___-___-___   expiration ___
EMAIL ___
PHONE   FAX

| 13# ISBN 2007 | TITLE | Qty | Retail | Total |
|---|---|---|---|---|
| 978-1-58197-884-1 | MTEL Biology 13 | | | |
| 978-1-58197-883-4 | MTEL Chemistry 12 | | | |
| 978-1-58197-875-9 | MTEL Communication and Literacy Skills 01 | | | |
| 978-1-58197-893-3 | MTEL Visual Art Sample Test 17 | | | |
| 978-1-58197-683-0 | MTEL Earth Science 14 | | | |
| 978-1-58197-879-7 | MTEL English 07 | | | |
| 978-1-58197-892-6 | MTEL Foundations of Reading 90 (requirement all El. Ed) | | | |
| 978-1-58197-887-2 | MTEL French 26 | | | |
| 978-1-58197-876-6 | MTEL General Curriculum (formerly Elementary) 03 | | | |
| 978-1-58197-877-3 | MTEL General Curriculum (formerly Elementary) 03 Sample Questions | | | |
| 978-1-58197-881-0 | MTEL General Science 10 | | | |
| 978-1-58197-878-0 | MTEL History 06 (Social Science) | | | |
| 978-1-58197-196-5 | MTEL Latin & Classical Humanities 15 | | | |
| 978-1-58197-880-3 | MTEL Mathematics 09 | | | |
| 978-1-58197-890-2 | MTEL Middle School Humanities 50 | | | |
| 978-1-58197-889-6 | MTEL Middle School Mathematics 47 | | | |
| 978-1-58197-891-9 | MTEL Middle School Mathematics-Science 51 | | | |
| 978-1-58197-886-5 | MTEL Physical Education 22 | | | |
| 978-1-58197-882-7 | MTEL Physics Sample Test 11 | | | |
| 978-1-58197-898-8 | MTEL Political Science/Political Philosophy 48 | | | |
| 978-1-58197-888-9 | MTEL Spanish 28 | | | |
| | | | SUBTOTAL | |
| | FOR PRODUCT PRICES VISIT WWW.XAMONLINE.COM | | Ship | $8.25 |
| | | | TOTAL | |

www.ingramcontent.com/pod-product-compliance
Lightning Source LLC
Chambersburg PA
CBHW080537300426
44111CB00017B/2771